T0323742

THE SECRET
FORMULA

THE SECRET FORMULA

How a Mathematical Duel
Inflamed Renaissance Italy and
Uncovered the Cubic Equation

FABIO TOSCANO

TRANSLATED BY ARTURO SANGALLI

PRINCETON UNIVERSITY PRESS
PRINCETON AND OXFORD

Published by Princeton University Press
41 William Street, Princeton, New Jersey 08540
6 Oxford Street, Woodstock, Oxfordshire OX20 1TR

press.princeton.edu

All Rights Reserved
ISBN 978-0-691-18367-1
ISBN (e-book) 978-0-691-20032-3

Library of Congress Control Number: 2020930672

British Library Cataloging-in-Publication Data is available

Editorial: Susannah Shoemaker
Production Editorial: Karen Carter
Text Design: Lorraine Doneker
Jacket/Cover Design: Carmina Alvarez
Production: Jacqueline Poirier
Publicity: Katie Lewis and Matthew Taylor
Copyeditor: Theresa Kornak

Jackert art: (left) Gerolamo Cardano, (right) Niccolò Tartaglia

The translation of this work has been funded by
Seps segretariato Europeo per le
Pubblicazioni Scientifiche

Via Val d'Aposa 7 – 40123 Bologna – Italy
HYPERLINK "mailto:seps@seps.it" seps@seps.it – www.seps.it

This book has been composed in Arno Pro and Bembo Std

Printed on acid-free paper. ∞

Printed in the United States of America

1 3 5 7 9 10 8 6 4 2

CONTENTS

PREFACE

The modern era of mathematics began in the first half of the sixteenth century with the discoveries made by a handful of outstanding Italian scholars. Their crucial contributions led to the awakening of the "great art," algebra, which had not experienced any significant progress in almost 3,000 years. These new results would soon be the source of numerous and fruitful developments in algebra, paving the way for the advance of the discipline and the important role it would play in mathematics and scientific knowledge in general.

It would be a mistake to assume that an event such as the reawakening and flourishing of algebra during the Italian Renaissance should be of interest only to experts in the field and historians of science. In fact, the incidents and situations related in this book not only represent a key period in the development of mathematics but also constitute a web of events remarkable enough to appeal to the general reader: situations rich in fictional flavor—intrigues, secrets, passionate debates—featuring some fascinating personages, both brilliant and bizarre, capable of excelling as much by virtue of their superior intellect as by their all-too-human weaknesses.

At the center of the story is a pivotal moment in the history of mathematics: the discovery of the general formula for the solution of the cubic equation and the subsequent developments, together with the captivating intertwinement of the lives of the main actors, among whom the names Niccolo Tartaglia and Gerolamo Cardano stand out.

All the facts had already appeared in print elsewhere, but we felt that they deserved to be told in the context of a story that would do justice to the protagonists, to their ingenuity and shortcomings, and that would reflect the extent of the subject without, however, exhausting it. We sought to condense the story into a reasonable number of pages by focusing on its most significant, interesting, and suggestive aspects. Often, by resorting to quotations, we let the personages speak for themselves, when their own words are more effective and compelling than any paraphrase. Given that the book was designed and written for readers without any specific technical knowledge, we tried to illustrate the mathematical questions and formulas using a language as plain and clear as possible.

I would like to conclude this brief preface by expressing my heartfelt gratitude to Paola Borgonovo for her meticulous and excellent revision work, which resulted in a greatly improved and better flowing text and at the same time allowed me to correct some imperfections. My sincere thanks to Silvia Tagliaferri for the proofreading and to Andrea Morando for creating the beautiful cover design.

Finally, I am indebted to my friend Paola Rigon, from the Classense Library in Ravenna, for the impressive amount of documentary material she obtained for me through national and international interlibrary loans, and to Martha Fabbri, editor of the Galapagos Series, for encouraging the publication of this book and supporting every stage of its production with unfailing dedication and boundless patience.

Needless to say, the responsibility for possible oversights, omissions, or errors is entirely mine.

THE SECRET
FORMULA

CHAPTER I

The Abbaco Master

The starless and ominous night came to an end. French troops surrounded the city ready to launch the attack, while the persistent tolling of the bells summoned the population to take up arms against the enemy. The date was February 19, 1512—Fat Thursday, a time to celebrate Carnival—and Brescia was about to witness one of its most tragic days.

The city that would later be known as "the Lioness of Italy" had already fallen under the French yoke in May 1509, after having been ruled for more than eighty years by the Most Serene Republic of Venice. Under the wise and liberal Venetian administration Brescia had become one of the most prosperous cities in Lombardy, only to fall prey to the arrogant and oppressive French rule, an increasing source of popular discontent. Seeking to restore the previous political order, some prominent citizens—Count Luigi Avogadro and other members of the local aristocracy among them—had thus begun to conspire against the French authorities, and later led the successful uprising of February 3, 1512. That day, with the help of Venetian soldiers—and thanks to the fact that most of the French troops had been redeployed to the siege of Bologna—Brescia chased away the foreign occupiers, forcing the remaining French soldiers to take refuge in the city fortress, known as the Castle.[1] The joy of the Brescians would, alas, prove short-lived.

Called back from Bologna, the twenty-three-year-old French general Gaston de Foix promptly regained Lombardy, and on February 17 reached with his army the walls of Brescia. In no time, the foreign forces surrounded the city. De Foix called on the insurgents to surrender, promising them the clemency of Louis XII, king of France, but his offer was curtly rejected. In the night of February 18, the young commander, together with some 500 lancers and 6,000 infantrymen ready for action, entered the castle, where those French soldiers who had escaped the uprising were still barricaded. The order to attack was given as soon as day broke, and shortly after this the carnage began.

The French garrison coming out of the castle pierced the first Venetian lines, and after joining the other units commanded by de Foix pushed toward the city center. Brescian fighters and Venetian troops offered a desperate resistance, but the mismatch of forces and the superior organization of the French soon overcame all defense efforts and the attack ended in a bloodbath. By the time the sun set on that gory Fat Thursday Brescia was back in French hands, its streets scattered with corpses.

But it was not over yet: an exemplary punishment was handed down to the city by the French troops, in the form of widespread destruction of unheard of ferocity. They looted and burned down houses, slaughtered men and children, and raped women in a maddening spree that lasted almost two days. Many leaders of the revolt were publicly and cruelly executed, further adding to the horror. Count Avogadro, the soul of the insurrection, "had his head savagely cut off and exhibited as a trophy on top of the People's Tower," and his remains "were hanged on the city gates, low enough so that dogs could feed on them."[2]

Not even the doors of places of worship stopped the invaders' rampage, as churches were ransacked and stripped of their treasures and valuable furnishings. To flee the violence, terrified citizens sought shelter in the cathedral, among them a poor widow and her two children: a twelve-year-old boy, Niccolo, and a younger girl. But the aggressors had no scruples in storming the temple to continue their depredation. During the assault, a French soldier targeted Niccolo and dealt him a blow to the head with his sabre. And then a second, and one more still: as a stream of blood started to gush from the boy's skull, he was hit twice in the face. Mercilessly, the blade cut through his mouth and teeth, fracturing his jaw and palate. He suffered five injuries in all; any of them could have been fatal, and yet Niccolo survived.

In the following weeks, unable to afford a doctor, the mother cared for her son by herself. Niccolo was incapable of speaking or eating, other than swallowing with great effort some liquid foods, and he remained in such a condition for several months. Thanks to his mother's care, he finally recovered from his wounds. Later, as an adult, he would grow a thick beard to conceal the deep and permanent scars that disfigured his face. Little by little he also regained the ability to speak, but the injuries to his mouth had left him with a stammer. Because of this his playmates gave him a mocking nickname that he adopted as his last name, today written in golden letters in the history of mathematics: Tartaglia. [In Italian, *tartagliare* means "to stammer."]

Born in Brescia, probably in 1499, Niccolo Tartaglia was one of the sons of "Micheletto cavallaro," a humble postal courier who delivered mail on horseback and from whom the famous mathematician inherited his short stature, if nothing else.[3] In an autobiographical page of his *Quesiti et inventioni diverse* (first published in 1546),[4] Niccolo affectionately describes his father; and in a dialogue with Gabriele Tadino, Knight of Rhodes and Prior of Barletta,[5] he mentions the composition of his family of origin.

PRIOR: Tell me again, what was your father's name?

NICCOLO: His name was Michele (Michael). And because Nature did not endow him with an adequate height, he was known as Micheletto (Little Michael).

PRIOR: Certainly, if Nature was not prodigal with regard to your father's height, it has not been more generous with yours.

NICCOLO: And I'm glad for that, because being so short proves to me that I am really his son. Even if he left us—my brother, my two sisters, and myself—almost nothing except fond memories of him, I have heard from many people well acquainted with my father that he was a good person. And this is for me a source of greater joy than inheriting a fortune from a disreputable parent.

PRIOR: What was your father's occupation?

NICCOLO: My father possessed a horse, which he rode to deliver post at the service of notables of Brescia; letters from Brescia to Bergamo, Crema, Verona, and other such places.[6]

Niccolo then declares not to know his father's last name, adding that his father passed away when he was five or six years old, leaving his family in the most dire poverty.

PRIOR: What was your father's family name?

NICCOLO: God knows I do not know. I don't remember his family name or his first name, except that as a child I always heard him called Micheletto Cavallaro. He may have had some other name, but not to my knowledge. The reason is that my father died when I was about six, leaving our family—my brother (slightly older than me), my sister (younger than me), and my mother— without any financial means. We went through very hard times, of which I will spare you the details. Under such circumstances, inquiring about my father's family name was the least thing in my mind.[7]

However, years later Tartaglia will mention in his testament "Zuampiero Fontana" as his "legitimate carnal brother."[8] This prompted several historians to consider "Fontana" as Tartaglia's real last name, but in fact none of the attempts to verify this interpretation were conclusive.[9] What is certain, though, is that Tartaglia wished to adopt the surname "Tartaglia" as a reminder of the personal drama he suffered during the 1512 sack of Brescia—"as a good memory of such a disgrace of mine,"[10] he writes—and perhaps also to remember the devoted and tender care his mother had provided him in those days of pain and suffering.[11]

During his conversation with the prior, Tartaglia reveals a few details of his early training. In particular, we learn that between the ages of five and six, shortly before his father's death, he was sent to a "reading school" run by a teacher whose name he did not remember.[12] Later, when he was around fourteen, he went "voluntarily and for about fifteen days to the writing school of a teacher called Francesco," who taught him "how to write a, b, c, and so on up to k in a script called 'mercant-esca' [of merchants]." This was a cursive script used in several cities of northern Italy to write documents and commercial books in the vernacular; its principal characteristics were the roundness of the letters and the richness of the joins.[13] To the next, predictable question of the prior, "why only up to the letter k?" Niccolo replies:

> Because it was agreed that I would pay the teacher one-third of his fee at the beginning, another third after I had learned the letters up to k, and the rest upon having learned the whole alphabet. And I didn't have the money to honor the last part of the agreement. However, since I wished to learn, I procured some alphabets and examples of letters written by the teacher's hand and never went back, because from these I learned by myself. And from that day on I never went to another teacher, and my only company was that daughter of poverty called Industriousness. I have continually studied the works of departed men.[14]

In a historical period and social context in which, save for rare exceptions, free public instruction did not exist,[15] the young Tartaglia had to work doggedly in order to cope with his dire financial situation and acquire the desired educational training. He achieved this by teaching himself, and mathematics was one of his first subjects. In his last book, *General trattato di numeri, et misure* [General Treaty of Numbers and Measures], Niccolo recalls having started the study of the discipline in 1514 and making such rapid progress that he soon was able to improve the rule to extract arithmetic roots.[16]

On the whole, Tartaglia built and perfected his own scientific training through the study of the works of various masters of the past (the "departed men"): from the great Greek thinkers, in particular Euclid, Archimedes, and Apollonius, to the medieval and Latin authors. He was thus obliged to become proficient in what was at the time the universal language of the learned, Latin, an indispensable tool for gaining access to the scholarly texts and their treasure trove of knowledge.

From what can be gleaned from the few available autobiographic notes about his youth, Tartaglia lived in Brescia until the age of eighteen or nineteen. Between 1516 and 1518, after spending some time wandering around "young and bachelor" in Crema, Bergamo, and Milan,[17] he left his native town and settled in Verona, where he stayed until 1534. We ignore the reasons for his going there, but whatever the case, Niccolo kept pleasant memories of the city of the Scala family: "Not only was it my first home away from the nest in which I was born," he writes in his *General trattato*, "but it always nourished me, caressed me, and honoured me."[18]

In Verona, Tartaglia, now in his twenties, married Domenica, a woman fourteen years his elder and mother of Benvenuta, an eight-year-old girl. The couple would later have their own child, Margherita, in 1527. For a while, family duties prevented Niccolo from devoting as much time to his studies as he would have wished. He nevertheless acquired a certain reputation, in Verona and other places of northern Italy, in his new role teaching practical mathematics or, more precisely, as an abbaco master.

During the thirteenth century, many Italian cities experienced a thriving increase in commercial activity. With the development of trading companies and the expansion of international trade, the new merchants were confronted with having to run ever-larger companies, whose administration required novel and more complex accounting procedures. In addition, the pressing need to master certain calculation methods was a common concern of shopkeepers, craftsmen, artists, and architects—in short, of all those involved in buying and selling goods, who had therefore to deal with, for example, currency equivalence, conversion from one unit (of weight, length, or area) to another, evaluation of assets and profits, and calculation of interest.

To satisfy the demand for training, in Italy around the middle of the thirteenth century flourished so-called abbaco schools—"institutions perhaps unique in late-medieval and Renaissance Europe,"[19] similar to present-day trade schools—where students learned those elements of practical mathematics that future merchants and technicians would require. At the time, therefore, the term "abbaco" not only referred to the wooden board with carved grooves within which beads could be moved around that ancients used to carry out arithmetical calculations, but it also denoted "the accounting operations and problems related to commercial practice."[20]

Altogether different, on the other hand, was the mathematics studied in those days at university: an abstract and speculative mathematics, whose paradigm was the elegant geometry of Euclid, and which aimed almost exclusively at satisfying theoretical and philosophical interests. Surely, academic geometry also "offered useful techniques for measuring heights, widths and areas," but their main purpose was "a quest for harmony."[21] Similarly, the arithmetic taught at the medieval lecture halls sought to "find harmonious numerical relations" that could be of interest to other disciplines, such as metaphysics or theology. But the principles of order, unity, and harmony so dear to university mathematicians "did not help merchants solve the disorderly problems of money exchange."[22] It should be noted, though, that abbaco mathematics was not entirely foreign to the academic world: there is evidence that at the end of the fourteenth century lectures on mathematics "*di minor guisa*" (in a smaller manner) were given at the University of Bologna, addressing the solution of practical and utilitarian problems "related to the exercise of professions or to events of civil life."[23]

Originally established in Tuscany, abbaco schools—which could be public or private—spread to most Italian cities during the last decades of the thirteenth century, starting a pedagogical tradition that continued until the end of the sixteenth century. Throughout this period, these schools provided a kind of secondary level education, after a first cycle of elementary schooling during which pupils learned to read and write and acquired basic arithmetic skills. At the age of ten or eleven, they could normally choose to enter either "grammar school," to study Latin language and literature and logic, or abbaco school, where commercial mathematics was taught in the vernacular (whereas the language of instruction of academic mathematics was Latin). While grammar schools gave access to university, abbaco schools—after about two years of training, depending on the place and the requirements of pupils—led straight to an apprenticeship with a merchant or craftsman as preparation for the exercise of a trade. As historians of science Enrico Gamba and Vico Montebelli observe, "Abbaco schools were attended either by nobles whose interests lay in commerce and the political rewards that it was expected to provide, or by members of lower classes seeking to improve their social and economic status through the acquisition of a professional qualification." [24]

Grammar and abbaco schools were not necessarily parallel and alternative educational paths, as "grammar could also be chosen subsequent to abbaco learning."[25] A fitting example of this possibility is the schooling of the celebrated Florentine philosopher, writer, and politician Niccolo Machiavelli. In 1480, at the age of eleven, he attended an abbaco school in his native city for one year and ten months, after which he went on to grammar school.[26]

The first public abbaco teacher was probably the eminent Pisan mathematician Leonardo Fibonacci,[27] whose life straddled the twelfth and thirteenth centuries, and to whom the authorities of Pisa awarded in 1241 an annual stipend of twenty pounds as consultant in commercial arithmetic for the Tuscan Municipality and its officials.* The young Fibonacci followed his father Guglielmo, a notary who assisted Pisan merchants with customs procedures, to the Algerian port city of Bugia. It was there that he became familiar with the avant-garde mathematics of the Arabs, which he would later further study during his journeys to Egypt, Syria, Greece, Sicily, and Provence. The culmination of all those years of peregrinations and research was a 1202 manuscript, the *Liber abaci* (Book of Calculation). This fundamental treatise written in Latin,[28] together with his numerous other works, earned Fibonacci the reputation of "most important medieval mathematician in the West."[29]

Divided into fifteen chapters, the *Liber abaci* is a voluminous collection of the arithmetic and algebraic knowledge of the Arab world, enriched with the author's own original contributions. The first part of the book introduces the Indo-Arabic numerals (that is, the usual ten numerals from 0 to 9) and positional notation.[30] If these are today familiar notions, they were practically unknown in the Europe of the time, where the cumbersome and inefficient Roman numeral system was still being used. Fibonacci then presents in the new notation the algorithms

* See, for example, [103c], pp. 256–57.

for the four arithmetic operations—addition, subtraction, multiplication, and division—on integers and fractions. The second part of the book discusses a variety of topics, including commercial mathematics—problems on purchases, sales, exchanges, and currencies; extraction of square and cubic roots; the theory of geometric proportions; and finally algebra.[31]

Even if it was considerably advanced with respect to the mathematical knowledge prevailing in the West at the beginning of the thirteenth century, it took nearly one hundred years before the *Liber abaci* was fully understood and appreciated. In Italy, it was through the abbaco schools that the new ideas proliferated, notably with the production of mathematical manuscripts based on Fibonacci's book, the "abbaco treatises," which began in the late thirteenth century. Written in the dialect of the various Italian regions, these volumes presented essentially the contents of the *Liber abaci*[32] in an elementary way, leaving out its most abstract parts.

There are some three hundred extant handwritten abbaco treatises,[33] the majority of them from the fifteenth century, written mostly by abbaco teachers for their students, or more generally for the benefit of anyone "wishing to have a handbook to which to refer for the solution of everyday problems arising in commercial operations."[34] Some of the authors are nonprofessional mathematicians, bankers or merchants, and, occasionally, even artists, as in the case of a 1480 *Trattato d'abaco* written by the great Tuscan painter Piero Della Francesca.[35]

At the turn of the fifteenth century, after the invention of the printing press by the German typographer Johannes Gutenberg, the first printed abbaco books made an appearance, their number increasing during the following decades and retaining essentially the same characteristics as handwritten ones. The anonymous work known as *Aritmetica di Treviso*, published in 1478 in Treviso, as its title indicates, is not only the oldest printed abbaco treatise but also the first ever printed book on a mathematical topic.[36]

The abbaco books of the Middle Ages and the Renaissance were fundamentally different from present-day mathematics textbooks, both in style and structure. An introduction explaining rules and definitions was followed by a long series of problems, each accompanied by an extremely detailed solution procedure. All this was expressed in narrative terms, as mathematics had not yet developed a symbolic notation— which does not make for easy reading, even for those with some mathematical knowledge. Here is an example:

> A soldo of Provence is worth 40 denari of Pisa and a soldo imperiale is worth 32 of Pisa. Tell me how much will I have of these two monies mixed together for 200 lire of Pisa? Do it thus: add together 40 and 32 making 72 (denari), which are 6 soldi, and divide 200 lire by 6, which gives 33 lire and 6 soldi and 8 denari, and you will have this much of each of these two monies, that is 33 lire 6 soldi 8 denari for the said 200 lire of Pisa. And it has been done.*

Abbaco books and teachers showed how to solve particular problems by analyzing them step by step, and for each problem a specific solution technique was provided. They did not teach general solution methods, as is the case in modern mathematics education. Quite the contrary: for a minor variation in the structure of a problem a different solution procedure was often proposed, the one regarded as the most appropriate and efficient for that particular case. Paul Grendler, an American specialist in schooling in Renaissance Italy, explains:

> The abbaco book collected individual problems and their solutions for reference use. A teacher found in it the day's problems and solutions to teach; students copied down what the teacher explained. The slow, literary statement of the problem and solution may have helped the students to understand and remember. If he faithfully copied enough problems and solutions, he had his own abbaco book. When as an adult merchant, banker, or clerk he came across a problem that he could not solve, he looked into his student abbaco book to find an

---

* This example is taken from [56], p. 315. The problem requires the conversion of 200 lire into numerically equal amounts of the two monies. Readers who would like to check the solution should keep in mind that in the then prevailing monetary system, adopted in most of Europe and based on lire (singular, lira), soldi (singular, soldo), and denari (singular, denaro), one soldo was worth 12 denari, and one lira 20 soldi.

almost identical problem and applied its method. Joining problem and solution together made the learner's task easier.[37]

The primary goal of abbaco schools was essentially the solution of problems, to which students were led not by logical deduction from general theoretical principles but through memorization of typical cases and preestablished rules. Considered as an "inseparable unit of a statement and its solution procedure,"[38] each problem was for the student a reference, a model to be followed in order to unravel a strictly similar practical question. Therefore, as Gamba and Montebelli observed in the foregoing quotes, abbaco teaching instilled in the student "a mnemonic, analogical, and practical mental attitude rather than a logico-deductive one, well-suited to the exercise of commerce, where situations that required rapidly counting one's and other's money abounded."[39]

One of the principal skills required of abbaco students was precisely the rapid execution of written, mental, and hand calculations. After learning the Indo-Arabic numerals and the arithmetic operations on integers and fractions, students had to memorize long multiplication tables to be able to perform complex calculations fast; they were also expected to check the correctness of the results. Tartaglia himself recommended "to verify diligently every step and operation, not just once but two and three times, because to err is human."[40] For the rest, the curriculum of the abbaco schools, which can be inferred from abbaco treatises, was standard and included commercial arithmetic (buying and selling, costs, profits, interest, discounts, currency conversions, conversions from one unit of measure to another, exchanges, and creation of capital companies, among others), practical geometry (especially the calculation of areas and volumes of concrete objects), and bookkeeping.

Even if the abbaco tradition failed to directly produce original mathematical results of any significance, it is a fact—and not a marginal one—that in late Medieval and Renaissance Italy it notably increased mathematical literacy through the gradual adoption of the Indo-Arabic number system and the introduction of mathematics into many professional activities. It would be a mistake, however, to reduce abbaco mathematics solely to its practical dimension. In the abbaco treatises—and foremost in Fibonacci's *Liber abaci*—one can find plenty of problems with no immediate practical application that fall in the category of "recreational" mathematics (games, riddles, and curiosities), and even algebra.[41]

It was precisely in the field of algebra that the contribution of abbaco mathematics was crucial, by paving the way to the most important mathematical discovery of the sixteenth century, one that marked the first real step forward of mathematics with respect to the ancient knowledge of the Greeks and the Arabs. It was a historic turning point, which had as one of its main protagonists the abbaco master from Brescia whose story we interrupted at the point where he moved to Verona.

According to archival sources, as early as 1284 the City of Verona had established a public abbaco school, and the following year a certain Lotto, from Florence, was appointed as a teacher with an annual stipend of fifty Veronan lire and free lodging.[42] Between the fourteenth and fifteenth centuries abbaco teaching thrived in Verona thanks to the pedagogical efforts of other Tuscan teachers and local tutors, appointed by the podesta (the highest judicial and military magistrate) and remunerated by the Merchants Guild, a powerful and influential arts and crafts corporation.

Niccolo Tartaglia was certainly a public abbaco teacher in Verona at least since 1529, as confirmed by official records held in the Archives of the State of Verona,[43] but it is likely that his appointment dated back to 1521 or shortly after. Other official documents, from the early 1530s, indicate that Tartaglia held classes at the Mazzanti Palace, located near Piazza delle Erbe, and that his financial situation was modest. This is confirmed by the 1531 Tax Roll, where it is stated that the tax assessed to "*Nicolas brixiensis magister abbachi*" (Niccolo from Brescia, abbaco teacher) was the meagre sum of zero lire and six soldi.[44] On the other hand, low pay and the resulting economic hardship was the lot of many abbaco and grammar teachers all over the country, who were often forced to move from city to city in search of better working conditions.

To supplement their insufficient income and make the most of their technical skills, abbaco teachers carried out various professional activities besides teaching; these included consultancy work, account auditing, land surveying, commercial advising, and assisting architects and engineers. Tartaglia was no exception: he once acted as accounting expert in a judicial dispute involving gemstone merchants in Verona;[45] on another occasion he was given the task of verifying the tables used by bakers to determine the weight of one soldo of bread as a function of the price of flour, which had undergone a sudden increase during a period—going back to 1531—when the city was hit by a severe famine.[46]

On the strictly mathematical front, the numerous problems in algebra, arithmetic, and geometry posed to Tartaglia by the most disparate people from Verona and elsewhere were clear evidence of his reputation as a talented abbaco master. For his part, Niccolo was well aware of the importance of dialogue with other experts or curious amateurs for his own investigations, and he stressed the benefits of having such conversations: "The questions and problems posed by wise and judicious interlocutors often prompt us to consider many things, and be acquainted with countless others which, had not been by the query, we would never have learned or considered."[47]

People of every social standing—merchants, engineers, architects, humanists, churchmen—turned to Tartaglia for advice or help with mathematical problems. He was also approached by amateur mathematicians and other abbaco teachers wishing to challenge him. And this is not just a manner of speaking: contests in which mathematicians challenged one another, true scientific duels carried out in a way reminiscent of chivalry tournaments, were very much in vogue in Italy in those times. A mathematician or scholar would send another a list of problems to be solved in a given amount of time—the "challenge gauntlet"—after which the recipient would propose a further set of problems to his rival. Tradition required that in case of disagreement a public debate should be held in which the contenders would discuss the disputed problems and solutions in front of judges, notaries, government officials, and a large crowd of spectators. It was not unusual in those duels for tempers to flare, and personal abuse take the place of scientific argument. Admittedly, the stakes could be very high: the winner of a public mathematical duel—whoever had solved the largest number of problems—gained not only glory and prestige but possibly also a monetary prize, new fee-paying disciples, appointment (or confirmation) to a chair, a salary increase, and, often, well-paid professional commissions. The defeated contender's future career, on the other hand, risked being seriously compromised.

Sixteenth-century mathematical duels had a long history and some illustrious predecessors. One famous example is the debate over mathematical, physical, and philosophical beliefs that took place in Ravenna on Christmas Day of the year 980 and that opposed the French humanist monk and scholar Gerbert of Aurillac and the German philosopher Otrich von Magdeburg. A large crowd, notably including Holy Roman Emperor Otto II, witnessed the day-long debate, with the contenders becoming "increasingly frenetic" and "showing no signs of stopping" as the hours passed.[48] At a certain point the emperor himself interrupted the discussion "because it was getting late and the audience was tired."[49]

In the end, it was the Frenchman who came out victorious, and not long after was appointed by Otto II, abbot at Bobbio, a small town in northern Italy. In 999, Gerbert of Aurillac would become pope and take the name Sylvester II. He was the first French pope in history and one of the most learned men of his time.

During the first half of the thirteenth century several mathematical contests had as protagonist the great and often-remembered Leonardo Fibonacci. Among the most remarkable ones are those in Pisa that placed him in competition against Johannes of Palermo, a scholar at the imperial court of Holy Roman Emperor Frederick II, in the presence of the emperor. Fibonacci later published the solutions to some important problems in algebra and arithmetic proposed to him by his rival during these contests.

Medieval university life was marked by scholarly debates—not only on mathematical questions—in which academics were required to participate. Ettore Bortolotti, a mathematics historian, writes:

A 1474 directive ordered lecturers[50] at the University of Bologna to gather in the public square or under the porticos at the end of classes and therein engage in debates and challenge one another; some of the most famous ones, held at Square Santo Stefano, are still remembered today. These daily confrontations served as preparation for the real contests that took place on designated days in the presence of the entire academic body and before huge crowds, the contenders having to conform to rules set out in the statutes. Every lecturer [at the University of Bologna] was required to take part in these public disputes at least twice a year.[51]

The Bolognian debates "were followed with great deference and admiration," and they were so popular that often "there wasn't a hall big enough to accommodate everybody."[52]

An unwritten rule of Renaissance mathematical duels was that a challenger should not propose to his rival any problem he was not able to solve himself. This was precisely the reason for Tartaglia's skepticism and annoyance in receiving, in 1530, two problems from a certain Zuanne de Tonini da Coi,[53] a mathematics teacher from Brescia: "Find me a number that multiplied by its root plus 3 makes 5. Similarly, find me three numbers, such that the second is greater by 2 than the first and the third is also greater by 2 than the second; and such that the first multiplied by the second, and this product multiplied by the third, makes 1000."[54]

Tartaglia soon realized that the problems led to two algebraic equations that in modern notation are written $x^3 + 3x^2 = 5$ and $x^3 + 6x^2 + 8x = 1000$ (they are solved by finding the value of the unknown number x that makes them true).*

The cause of Tartaglia's puzzlement was simple: these are third-degree equations (also called *cubic* equations), which means that the unknown x is raised to the third power (or *cube*, $x^3 = x \times x \times x$), but at the time no general formula for solving all equations of this type was known. Actually, certain very particular cases of the equation could be

---

* For the interested reader: in the first problem, the equation is obtained by letting x^2 be the unknown number and x its square root; then, the expression $x^2 (x + 3) = 5$ follows from the statement of the problem, and the final equation is arrived at by algebraic manipulation. The equation in the second problem results from denoting the unknown numbers by x, $x + 2$, and $x + 4$, respectively, so that we have $x(x + 2)(x + 4) = 1000$, from which the final equation can be obtained.

solved by approximation methods, but Tonini da Coi's were not among these. A few decades earlier, in his monumental and influential *Summa* published in 1494,[55] the famous Tuscan mathematician Luca Pacioli—a Franciscan monk who taught in several Italian cities—had considered it "impossible" to solve the general equation by means of an algebraic formula with the algorithmic tools then available.[56] According to Pacioli, "art"—that is, algebra—could not have yet "formed" general rules for solving this kind of equation "except by feeling one's way in the dark, [. . .] in some particular cases."[57]

Was it then possible for such a not particularly gifted mathematician as Zuanne de Tonini da Coi to have suddenly pushed back the boundaries of algebraic knowledge when the ancient masters had failed? Tartaglia refused to believe it, even for a second. With his characteristic stubborn and caustic disposition, he replied to Messer Zuanne* in rather polemic terms, accusing him of bragging and ignoring Pacioli's highly respected opinion:

> Messer Zuanne, you have sent me these two questions of yours as something impossible to solve, or at least as being unknown to you, since proceeding by algebra the first leads to a cube plus 3 cenno equal to 5 $[x^3 + 3x^2 = 5]$ and the second to a cube plus 6 cenno plus 8 things equal to 1000 $[x^3 + 6x^2 + 8x = 1000]$.[†] These chapters[58] have until now been considered as impossible to solve with a general rule by Fra Luca and others. You believe with these questions to place yourself above me and appear as a great mathematician, as I heard you have done with all the other professors of this science in Brescia, who fearing your questions do not dare speak to you. Perhaps they understand this science better than you without knowing all the answers, but believing you do, they concede everything to you.[59]

---

* The word Messer is an archaic title of courtesy prefixed to the first name (sometimes with the surname also) of an Italian man. It is more or less equivalent to the modern Italian Signor, which in turn is equivalent to the English Mr.

† In the algebraic language of the time, which was nonsymbolic, the term "thing" [cosa, in Italian] designated the unknown x, whereas "cenno" [censo] indicated the square of the unknown, that is, x^2. These terms later disappeared from the mathematical vocabulary, while "cube" [cubo], to express x^3, is still used today.

It appears then that this was not the first time Tonini da Coi had tried to impress fellow mathematicians, who, unaware of his inclination for boasting, might have been intimidated by his questions. But Tartaglia proved too clever for him, and da Coi received an abrupt and harsh rebuke. Niccolo finished his letter by declaring himself ready to bet that his challenger was incapable of solving the problems he had so daringly proposed:

> Hence, to correct the vain opinion you have of yourself and induce you to seek honors through knowledge rather than by posing questions you know nothing about, I reply to you that asking others for solutions you cannot find yourself should make you blush. And to prove to you how certain I am of this, I am willing to wager you ten ducats against five that you will not be able to solve with a general rule the two cases you proposed to me.[60]

His bluff being called, Messer Zuanne nevertheless took the blow, and in a subsequent letter asked Tartaglia if he, too—as his reply suggested—considered it impossible to find an algebraic formula to solve the cubic equations in question.

Niccolo replied with an unexpected admission: "I am not saying that such cases are impossible. On the contrary, for the first one, that of cube and cenno equal to number [$x^3 + 3x^2 = 5$ in modern notation] I am convinced to have found the general rule, but for the time being I prefer not to reveal it for several reasons [. . .]"[61]

Not to reveal it? Why would Tartaglia wish to keep secret a discovery of historic significance, which would be certain to bring him enormous prestige in mathematical circles? What could be his reasons for choosing silence? Had he really discovered a "general rule" for solving those types of cubic equations, or was he, too, indulging in unfounded boasting?

Only Niccolo knew the truth. In any case, in those times it was customary among mathematicians to keep their methods and results secret as long as possible, either for fear that once revealed, students or potential clients would no longer require their services, or for the purpose of eventually using them to their advantage in duels with other scholars. Seen in this light, Tartaglia's attitude merely conformed to the customs of his time. Besides, in his second letter in reply to Tonini da Coi, Niccolo admitted to not having yet found a general formula for solving the second equation proposed by Zuanne, adding, however, that it should not be impossible to obtain: "[...] as regards the second case, that of cube and cenno and thing equal to number [the equation $x^3 + 6x^2 + 8x = 1000$], I must admit not having been able until now to find a general rule, but I am not saying that one is impossible to find, even if none has so far been discovered."[62] The letter ends with yet another harsh scolding addressed to Tonini da Coi, and a repeated invitation to "blush" for having flaunted knowledge he did not possess.

And yet, it was precisely thanks to the questions posed by that braggart Messer Zuanne in 1530 that Niccolo began to turn his attention to those elusive cubic equations with the intention of finally cracking their mystery—an event that mathematics had been awaiting for millennia.

CHAPTER 2

The Rule of the Thing

Its code name is TM 75 G 1693. Made up of clay, its round shape fits in the palm of a hand. What makes this 4,500-year-old artefact special are the inscriptions on it, believed to be the oldest recorded equations. "The most ancient document with possible algebraic significance,"[1] writes mathematics historian Silvio Maracchia.

This extraordinary archaeological find comes from the magnificent civilization of Ebla, a city located in what is now northern Syria and whose vestiges were brought to light by a 1964 Italian expedition headed by Paolo Matthiae, an archaeologist at La Sapienza University in Rome. In 1975, the royal archives were discovered: a treasure trove of more than 17,000 clay tablets dating back to 2500 BCE, inscribed in cuneiform script, mostly in a Semitic language known as Eblaite, and containing economic, administrative, religious, historic, judicial, and literary texts.[2] Among these innumerable tablets was also TM 75 G 1693, with its undeniably mathematical inscriptions, but whose true meaning remained to be deciphered.

It did not take long, thanks to the translation work of the Orientalist Giovanni Pettinato, to learn that the tablet in question had been written by a scribe named Jsma-Ja, from the Sumerian city of Kish, in Mesopotamia. We may imagine him, perhaps "the first mathematician in history,"[3] as a sort of foreign teacher invited to Ebla, on account of his reputation, to lecture to local students.[4] This seems to be confirmed by the tablet's contents, deciphered by the Italian scholar as follows:

600 gal

3,600 gal

36,000 gal

360,000 gal

$360,000 \times 6$ gal

not solved

This text[5] appears to be an exercise assigned by Jsma-Ja to his students. But what kind of exercise?

The key to understanding the problem posed by the scribe lies in the word *gal*, which literally means "big." However, to discover the meaning of the text it is necessary to interpret this word in a more specific and relevant way, a challenge that several mathematicians and mathematics historians took up. Among these were the Italians Tullio Viola and Isabella Vino,[6] and so far their interpretation is considered by many scholars as the most reliable, even if it is not the only plausible explanation. Based on a careful philological analysis and comparison with other inscriptions, the two mathematicians concluded that the word *gal* should be understood as a multiplicative factor of 60: that is, Jsma-Jal asked his students to find those numbers that multiplied by 60 give 600, 3,600, 36,000, 360,000, and $360,000 \times 6$, respectively. In other words, the students had to find the solutions of the following algebraic equations expressed in modern mathematical symbols:

$$60x = 600$$
$$60x = 3,600$$
$$60x = 36,000$$
$$60x = 360,000$$
$$60x = 360,000 \times 6$$

Each of these equations is said to be of first degree because the unknown x occurs raised to the first power, that is, not raised at all (since $x^1 = x$). First-degree equations are the simplest to solve. It is easy to see that $x = 10$ is the solution to the first equation proposed by Jsma-Ja, given that $60 \times 10 = 600$. The four other equations involve larger numbers but none of them seem particularly difficult; it is readily found that the values of x that satisfy them are, respectively, 60, 600, 6,000, and 36,000.

All in all, the questions asked by Jsma-Ja were fairly simple, and yet, as tablet TM 75 G 1693 clearly informs us, they were not answered. Of course, there is nothing to prevent us from presuming that the students were not very clever, or lazy, or that the scribe was an ineffective teacher. Pettinato observes, however, that from the start, a possible stumbling block for the students was the fact that they had to reason in the Sumerian number system (sexagesimal, or base 60) used by Jsma-Ja, considerably different from the local one (decimal, or base 10) and therefore "not congenial to them."[7]

Whatever the case, those certainly not daunted by first-degree equations were the mathematicians of an ancient people that have for centuries fascinated scholars and the public at large: the Egyptians.

The scarce documentary sources available on Egyptian mathematics consist essentially of some papyrus rolls that have survived the injuries of time. Among these, the two most important by far are the *Moscow papyrus*, dating from ca. 1890 BCE and kept in the Pushkin Museum of Fine Arts in the Russian capital, and especially the *Rhind papyrus*, two and a half centuries more recent, housed in the British Museum in London (except for a few small fragments kept in New York's Brooklyn Museum).

The Moscow papyrus, some five and a half meters long and about eight centimeters wide, it is also known as the Golenishchev papyrus, after the eminent Russian Egyptologist Vladimir Golenishchev, who bought it in 1893. Some twenty years later, in 1912, he decided to donate it to the state together with his vast collection of Egyptian antiquities in exchange for an annuity. Unfortunately for him, after the October 1917 Revolution the payments stopped.

The Moscow papyrus, written by an unknown scribe in hieratic, a simplified form of hieroglyphic script, features twenty-five mathematical problems, not all of them legible or easily interpreted. Among the most significant ones, problem 14 represents, in the experts' opinion, the highest achievement of Egyptian geometry: the correct calculation of the volume of a truncated square-base pyramid. This is the first historical evidence of such a result—and the only one in the ancient world.

The problems proposed in the Rhind papyrus are far more numerous—eighty-seven in all—and form the basis of most of our knowledge of Egyptian mathematics. This remarkable document is named after the Scottish lawyer, antiquarian, and egyptologist Alexander Henry Rhind, who purchased it in 1858 in Luxor, a famous city on the left bank of the River Nile, built on the site of the ancient and splendid Thebes. Almost as long as the Moscow papyrus—slightly less than five and a half meters—but about four times as wide, some thirty centimeters, the Rhind papyrus was the work of a scribe named Ahmes, at around 1650 BCE. By Ahmes' own admission, he copied it from an original text a couple of centuries older.

Written in hieratic script, the Rhind papyrus offers a broad and varied portrait of Egyptian mathematics, which although far from attaining summits of excellence is nevertheless of great historical interest. Admittedly, the papyrus' opening title—"The entrance into the knowledge of all existing things and all obscure secrets"[8]—is definitely overambitious given the nature of the topics treated in the text, for the most part concerning practical questions and never developing an abstract, systematic general theory. The presentation of two division tables is followed by the statement and solution of a number of arithmetic and geometric problems of delightfully empirical character, often related to everyday-life situations: equitable distribution of loaves of bread among a certain number of individuals; conversion of a given amount of barley into beer; calculation of the slope of a pyramid; payment of wages of construction workers or a temple's staff.

Not all the eighty-seven problems in the Rhind papyrus address practical questions. Problem 79, for instance, is a true mathematical riddle, which may be stated as follows:

In each of seven houses live seven cats; each cat kills seven mice; each mouse had eaten seven ears of wheat; each ear produces seven *hekat* of grain.[9] What number is the total of all?[10]

Alas, this mathematical riddle of the ancient Egyptians traveled across time (the answer is $7 + 7^2 + 7^3 + 7^4 + 7^5 = 19{,}607$). Disregarding the variations it underwent over time and space, it reemerged in 1202 in Leonardo Fibonacci's *Liber abaci*:

Seven old ladies go to Rome; each of them has seven mules; each mule carries seven bags; in each bag there are seven loaves; each loaf has seven knives; each knife has seven sheaths. You are to find the sum of all these.[11]

The riddle also makes a modern appearance in a popular nursery rhyme:

On the road to Camogli
travels a man with seven wives;
each wife carries seven bags,
in each bag there are seven cats,
and each cat nourishes seven kittens.
Cats, kittens, bags, and wives,
How many of them are going to Camogli?[12]

Other problems in the Rhind papyrus with no immediate practical application are of particular interest for the history of algebra, for they can be solved with first-degree equations. These are the so-called "*aha* problems," from the Egyptian word *aha*, which means "heap" or "pile." This term indicated the numerical quantity to be calculated: in other words, *aha* was the name with which the Egyptians designated the unknown of an equation. Nowadays, the unknown is represented by the letter x (or some other letter toward the end of the alphabet), a convention that was originated in the first half of the seventeenth century by the great French mathematician and philosopher René Descartes.[13]

A typical *aha* problem from the Rhind papyrus is number 26, in which it is asked to find the value of a "heap" knowing that the sum of the heap and one-fourth of it equals 15. The scribe Ahmes states and solves the problem using only words (algebraic symbols did not exist in Egyptian mathematics); in modern notation the question would be expressed as follows:

$$x + \tfrac{1}{4}x = 15$$

To solve this equation, and in general any first-degree equation, a few simple operations are enough. To begin with, we can multiply each term in the equation by 4, so as to get rid of the fraction (four times one-quarter of x equals x). On the other hand, algebra teaches us that performing a given operation on all the terms of an equation results in a new equation *equivalent* to the original one: both equations have the same solutions, that is, they are satisfied by the same values of x. Therefore, besides simplifying the equation, our multiplication by 4 is perfectly correct and leads to the expression:

$$4x + x = 60$$

The problem has now become much easier. We begin by adding four times x to x, which gives us five times x, and therefore we obtain

$$5x = 60$$

Finally, dividing both members of this last equality by 5, the left side becomes x and the right 12. Conclusion: $x = 12$ is the sought-after solution of the equation.

It is more or less in this way that a present-day high school student would solve it, as well as any other similar first-degree equation. Altogether different is the procedure described by Ahmes, the so-called "false position" method. To understand this technique, which may appear to us a little contrived, it is useful to see how the Egyptian scribe applies it to the equation

$$x + \tfrac{1}{4}x = 15.$$

The unknown x is given an arbitrary value, with which the operations on the left of the equal sign are carried out. In this case, Ahmes sets for convenience $x = 4$, so that after the substitution of x by 4 he can get rid of the fraction ($\tfrac{1}{4} \times 4 = 1$). This gives 5 and not 15, as required by the problem. But writing $5 = 15$ does not make sense; hence $x = 4$ is a "false position"; that is, it is not the correct solution. Nonetheless, Ahmes goes on, resorting to the theory of proportions: since 15 is the triple of 5, then the true x should be the triple of 4, i.e., 12, which, as we have seen, is the right solution.

Practically all first-degree equations studied by Egyptian mathematicians—at least those we know of—are solved by the false position method. A few examples of solutions using an alternative method comparable to modern techniques also appear in the Rhind and Moscow papyruses, but we are talking here of very, very few exceptions to the rule. Curiously, the false position method—refined and adapted to more complex situations in the course of time—was used by algebraists almost until the end of the nineteenth century, and its teaching "was still recommended in the Austrian gymnasiums as late as 1884."[14]

Some fragments of another document, the *Berlin papyrus*—from about 1800 BCE and kept today in the Ägyptisches Museum of the German capital—indicate that the ancient Egyptians also knew how to solve the simplest second-degree (or *quadratic*) equations, the *pure equations*, in which the unknown occurs raised only to the second power (i.e., squared), such as, for example, $x^2 = 9$. The solution to this simple equation is the square root of 9 (in symbols $\sqrt{9}$), where the square root of a given number is the number that squared gives the original number. In this case, it is straightforward to infer that $\sqrt{9} = 3$, since $3 \times 3 = 9$.

Besides 3, the given equation has another, less expected solution: −3. Indeed, according to our familiar arithmetic rules, the product of two negative numbers is positive; and because this is also true when a number is multiplied by itself, we have $(-3)^2 = (-3) \times (-3) = 9$. So that in fact the quadratic equation $x^2 = 9$ has two solutions, but the Egyptians considered only the first one because they did not possess the notion of negative number.

Not all cases have such a simple solution; some lead to less comfortable situations from a numerical point of view: the equation $x^2 = 5$, for example. Reasoning as in the previous equation, the two solutions are $x = \sqrt{5}$ and $x = -\sqrt{5}$. Now, $\sqrt{5}$ is not a whole number (or integer) but what mathematicians call an *irrational number*, or a decimal number in which the decimal point is followed by an infinite sequence of digits with no discernible order (the most famous irrational number is probably π, the Greek letter pi, usually approximated as 3.14). Hence, if we wish to know the value of $\sqrt{5}$, we have to make do with an approximate decimal expression, for example, 2.236067978, with nine digits after the decimal point.

Egyptian mathematicians treated what, *viewed from a modern perspective and using modern terminology*, we identify as pure quadratic equations. But they did not consider the "mixed" ones, where both x^2 and x occur, as, for example, $x^2 + 4x = 21$. To find those who tackled and solved such equations in early antiquity we have to turn our attention to the "land between two rivers," Mesopotamia, between the Tigris and the Euphrates.

Our knowledge of Mesopotamian mathematics comes from more than 700 clay tablets—including our old acquaintance TM 75 G 1693—written in cuneiform script and dated over a long period of time, spanning from the third millennium to the second century BCE. The vast majority of these documents go back to Babylonian times, in particular to around 1800–1600 BCE.

These mathematical tablets contain either calculation tables—for multiplication, division, powers, square roots, and cubic roots—or statements of problems in arithmetic, algebra, and geometry, generally connected to practical questions and not always accompanied by their solution. Whenever included, solutions apply only to the particular case stated in the problem, with no hint of a generalization, formula, or theorem.

"The Mesopotamian populations," observes mathematics historian Livia Giacardi, "did not consider mathematics to be a speculative and abstract activity requiring logic or rigor, but a social product generated by the needs of a continually expanding society. Babylonians lacked the scientific attitude characteristic of the Greek culture, and the reason for this lies in part in the context in which mathematics develops. Mathematics is, at its origins, an instrument of knowledge and power. It is born and develops in the temples as an essential tool for the administration of the city (construction of buildings and canals, tax collection, estate distribution, interest calculation, etc.), the measure of time, and for the regulation of agricultural and commercial activities, making it an integral part of the cultural knowledge of any future state official. The surviving mathematical texts appear to have been written mainly for pedagogical purposes."[15]

"The concrete character of the problems,"[16] the scholar continues, is evidence of their didactical function. However, the fact that problems in the tablets are classified according to the type of solution is a "symptom of a certain level of awareness of the notion of generality"[17] on the part of Babylonian mathematicians, despite the absence of demonstrations in their texts.

While being careful—as the renowned Austrian historian Otto Neugebauer warned—to not misunderstand or overvalue its achievements,[18] we can affirm that Mesopotamian mathematics succeeded in obtaining results of significant importance, indicative of a complexity level considerably higher than that of Egyptian mathematics. This is even more manifest in the domain of algebra, where Babylonians went well beyond simple first-degree equations: they also successfully addressed complete second-degree equations, third-degree equations, and some equations of higher degree reducible to those of second and third degree.

As is the case with Egyptian papyruses, the mathematical tablets of ancient Mesopotamia do not reveal the slightest use of algebraic symbolism; problems and solution procedures (when present) are expressed in words; unknowns are designated by terms borrowed from geometry, such as "length" (*uš*), "width" (*sag*), "area" (*a-šà*), or "volume" (*sahar*). These terms, however, must have been used in a very abstract sense; this can be inferred, observes mathematics historian Carl Boyer, from the fact that "the Babylonians had no qualms about adding a "length" to an "area" or an "area" to a "volume,"[19] operations devoid of geometric meaning and therefore not related to practical applications of measurements.

Babylonian mathematicians considered first-degree equations rather trivial and hence unworthy of much attention, since—as the tablets we possess indicate—they often wrote the answers directly, without showing the intermediate steps. The case of second-degree equations is different: solutions are explicitly presented, each one in reference to a particular problem, and take the form, if we can put it that way, of a "recipe." But nothing general, never a formula or a demonstration, "even if there can be no doubt that the Babylonians knew general calculation rules."[20]

Let us take a look at one of the problems inscribed in a 4,000-year-old clay tablet: BM 13901, kept in the British Museum (as indicated by "BM"). The text, composed by an anonymous scribe, asks to find the side of a square whose area (side × side = side²) minus the side is equal to 870; in modern algebraic symbols, it is equivalent to finding the solution of the quadratic equation

$$x^2 - x = 870,$$

where the unknown x denotes the side of the square. Here is the solution strategy proposed by the scribe:

Divide 1 into two parts: 0.5. Multiply 0.5 and 0.5: 0.25. You add to 870, and 870.25 has the root 29.5. You add to 29.5 the 0.5 which you have multiplied by itself: 30, and this is the side of the square.[21]

The "recipe" prescribes the solution $x = 30$, which is correct, as can be easily verified by substituting x by 30 in the above equation: 30^2 ($= 30 \times 30$) equals 900, and 900 minus 30 gives 870. Translating the scribe's instructions in usual arithmetic notation we obtain

$$x = \sqrt{(0.5)^2 + 870} + 0.5.$$

The equation $x^2 - x = 870$ is a particular case of a general quadratic equation of the form $x^2 - px = q$, where the letters p and q, the *coefficients* of the equation, represent any two positive numbers (in this case, $p = 1$ and $q = 870$). We can thus generalize the scribe's prescription and deduce the formula

$$x = \sqrt{\left(\frac{p}{2}\right)^2 + q} + \frac{p}{2}.$$

Now, this "recipe" is equivalent to the formula for the quadratic equation that appears in every mathematics textbook and today's students learn at school.[22]

The only significant difference is that we now couple this formula with a second, identical one, except that the square root symbol is preceded by a minus sign. This means that the equation $x^2 - x = 870$ has another solution besides $x = 30$, namely $x = -29$, as can be verified by replacing p and q by 1 and 870, respectively, in the above formula, and remembering that the value of the square root must be subtracted as well as added.

In the Mesopotamian texts, though, a possible negative solution is never mentioned, for the simple reason that the Babylonians—like the Egyptians—did not possess the notion of negative number (besides, in their problems, such as the one we have just discussed, the unknowns generally represented lengths, which are expressed only as positive numbers). Even so, there is no doubt that four millennia ago Babylonian mathematicians already knew how to solve quadratic equations. The question is: how did they get there?

Let us say, to begin with, that there is no definite answer to this question. We have already noticed the absence of the slightest trace of any mathematical demonstration in the Mesopotamian tablets. Nonetheless, as a result of multiple textual analyses experts concluded that the Babylonians solved quadratic equations through the use of an ingenious geometric method.

We shall illustrate it with a problem much like the one in tablet BM 13901 but better suited to our purpose. Suppose we wanted to find the side of a square whose area, added to four times the side, equals 21. In modern algebraic symbolism, we must solve the equation $x^2 + 4x = 21$, similar to the one recently discussed, where x indicates the side of the square. Reproducing the reasoning of the Babylonian mathematicians—such as it has been reconstructed—let us draw a square of side x and on two consecutive sides construct two identical rectangles of sides x (the long side) and 2 (the short side).

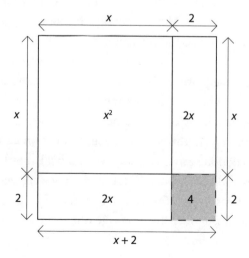

The area of each rectangle equals $2x$ (the product of the short and the long sides, or 2 and x), and therefore their combined area is equal to $4x$. Hence, the expression on the left in the equation we are trying to solve, $x^2 + 4x$, represents the area of the square (x times $x = x^2$) plus that of the two rectangles; all these figures taken together form another figure whose area is known to be equal to 21. This latter figure, as shown in the drawing, is "a shape like a square with a corner missing."[23] We can make it into a complete square by adding a small square of side 2. Inserting the latter, we obtain a new, large square whose area is equal to the sum of the areas of the initial square, the two rectangles, and the small square ($2 \times 2 = 4$), that is $x^2 + 4x + 4$. Since we know that the area of the quasi-square is 21, it follows that the area of the large square must be $21 + 4 = 25$. All this implies that, in algebraic terms, the problem's original equation can be written $x^2 + 4x + 4 = 25$. Now, for the last step: since the side of the large square equals $x + 2$, its area is $(x+2)^2$, and we can therefore express the problem's equation as $(x+2)^2 = 25$, leading to the first-degree equation $x + 2 = 5$, from which easily—and finally—we obtain the (positive) solution of the problem $x = 3$.[24]

Skillful and clever in their treatment of quadratic equations, Babylonian mathematicians also obtained significant results—even if very limited—with third-degree, or cubic, equations. In particular, they knew how to solve pure cubic equations, those in which the unknown occurs raised only to the third power (or "cube"), such as for instance $x^3 = 8$. In mathematical terminology, the solution of this equation is the "cubic root" of 8, written $x = \sqrt[3]{8}$. More generally, the cubic root of a number is the number whose cube is the given number, so $\sqrt[3]{8} = 2$, since $2 \times 2 \times 2 = 8$.

Just as with square roots, cubic roots are not always so easily calculated. When solving the equation $x^3 = 7$, for example, we run into $x = \sqrt[3]{7}$, an irrational number, whose approximation with nine digits after the decimal point is 1.912931183. Babylonian mathematicians got around situations like this quite nicely. In this specific case, to solve an equation such as $x^3 = a$ (where a denotes any given positive number), they looked up their tables of cubes and cubic roots; if the value they were searching for did not appear, they resorted to a special technique of numerical analysis— known today as "linear interpolation"—to obtain an approximate answer.

In an analogous way, Babylonians could also solve mixed cubic equations of the form $x^3 + x^2 = a$ through the use of tables listing the values of the expression $n^3 + n^2$ for whole numbers from 1 to 30. They even succeeded in solving more general cubic equations after having reduced them to the form just shown with smart manipulations.

Despite lacking any kind of mathematical symbolism, Mesopotamian algebra reached a remarkable level of achievement, thanks to a truly amazing technical and conceptual versatility. Equally remarkable is the fact that during many centuries the glorious and brilliant Greek mathematicians ignored the Babylonian algebraic tradition almost entirely.[25] As for the reasons for this, "we are left with hypotheses,"[26] warns historian Silvio Maracchia; in any case, Greek mathematicians devoted their attention almost exclusively to geometry and for a long time neglected algebra. It is true that in Greek mathematics we find "geometric constructions equivalent to the solution of equations or to certain algebraic transformations,"[27] so much so that the expression "geometric algebra" was coined to describe precisely that field of study, but to speak of "algebra" would be incorrect. This discipline was again in the limelight only at around the third century CE, with the works of one of the last Hellenic masters: Diophantus of Alexandria.

Almost nothing is known of Diophantus' life, not even the exact period in which he lived. On the basis of certain indications in his works and those of others, mathematical historians infer that he was active in Egypt's Alexandria at around CE 250. The only information on the mathematician's life comes from an epigram that according to tradition was used as the epitaph on his tomb. The text appears in Book XIV of *Antologia Palatina*, in which the compiler (Metrodorus of Bizance, fifth to sixth century CE) lists 150 epigrams, many of which are mathematical riddles.

During one-sixth of his life he was a child; after another one-twelfth he began to sport a beard. After a further seventh of his life, he took a wife, who five years later gave him a beautiful son. Unfortunately, when he reached half of his father's lifespan, the son suddenly passed away. The surviving father mourned his son for four years and then followed him.[28]

This epigram is clearly an algebraic problem. Reformulating it as a first-degree equation, where the unknown x denotes the age of Diophantus at his death, we obtain:

$$\frac{1}{6}x + \frac{1}{12}x + \frac{1}{7}x + 5 + \frac{1}{2}x + 4 = x$$

Assuming that Metrodorus' data were correct, we can conclude from the above equation (as the reader is invited to verify . . .) that Diophantus lived until the age of eighty-four, a venerable age for the time.

Among Diophantus' known works, the most important is unquestionably his *Arithmetica*, a treatise originally divided into thirteen books, ten of which survived: six of these come from a thirteenth-century Greek manuscript today housed in the National Library of Spain in Madrid; the other four (on whose authenticity experts diverge) were discovered only in 1968 in Mashhad, a city in northeast Iran. It is a handwritten Arabic translation by Qusta ibn Luqa, an Arab Christian physician and scientist who lived from 820 to 912. Refined, original, dotted with a myriad of mathematical virtuosities, *Arithmetica* is a vast collection of some 200 problems leading to equations of various kinds.[29] One of its novelties is the formulation of the problems, always rigorously abstract; only in a subsequent stage does the author provide numerical data, followed by a solution of the problem through algebraic and arithmetic procedures independent of any geometric interpretation. In some problems Diophantus' goes so far as to work with powers higher than three, which "the classical Greeks could not and would not consider"[30] given their lack of geometric meaning.

Another characteristic feature of *Arithmetica* is the use of abbreviations and symbols to express the unknown, its powers, and certain operations. It is the first instance of an extensive use of algebraic symbolism in the history of mathematics; even if incomplete and rudimentary, it counts among "one of Diophantus' major advancements."[31] Conventionally, in the historical development of algebra three characterizations of the discipline can be distinguished in relation to the *symbolism* employed: *rhetorical algebra*, in which everything (problems, solutions, demonstrations) is written in full sentences; *syncopated algebra*, in which both words—sometimes abbreviated—and symbols are used; and *symbolic algebra*, where only symbols are employed. Syncopated algebra was introduced by Diophantus and constituted a stepping stone toward symbolic algebra, which took hold on a permanent basis at the end of the sixteenth century, following French mathematician François Viète's notational innovations.[32] "These three stages have no sharp chronological boundaries,"[33] observe historians Raffaella Franci and Laura Toti Rigatelli. We have already seen evidence of this, as the algebra of the abbaco teachers of the Italian Middle Ages and Renaissance was still entirely rhetorical, and we shall find more examples in the coming pages.

In *Arithmetica*, Diophantus solves various kinds of equations without proposing any general formula; he presents only particular cases, although from his examples it is possible to extract the rules he had certainly mastered and followed. He notably tackles first- and second-degree equations but does not provide the explicit solution formula for the latter, "by now well known by all readers to whom his work was addressed."[34] In one of the problems, Diophantus arrives at the third-degree equation $x^3 + x = 4x^2 + 4$, of which he gives the correct solution $x = 4$ with no indication whatsoever of the method employed to find it. One plausible hypothesis is that he might have exploited a special feature of this equation: a *common factor factorization* of each member leads to $x(x^2 + 1) = 4(x^2 + 1)$. From this, since the parentheses on each side contain the same expression, the result easily follows.

Because of his works, Diophantus is often referred to as "the father of algebra." In fact, another major protagonist of the history of mathematics claimed this title—perhaps rightfully so. In order to find him, we have to travel in space and time toward a new horizon: the Islamic Middle Ages.

Already in 750, a little more than a century after the death of the founder of Islam, the prophet Muhammad, on June 8, 632, at Medina (in what today is Saudi Arabia), Arab domination spread over a vast territory, from Spain to the borders of India. Until then, the Arab conquerors, engaged in their unstoppable expansionist campaign, had manifested no appreciable intellectual interest. But from the middle of the eighth century they showed an increasing inclination for the arts, philosophy, and science. As a result, they allowed their own scholars to rescue from oblivion a considerable part of ancient and classical knowledge and make their own original and significant contributions. After 750, scholars from Syria, Persia, and Mesopotamia were summoned to Baghdad, the political center of the new Muslim Abbasid dynasty. The thriving city soon became the new cultural capital of the world of that time, taking over that distinction from Egypt's Alexandria. At around 830, the caliph al-Ma'mun founded the House of Wisdom (*Bayt al-Hikma*, in Arabic), a cultural institution housing a huge library—comparable to that of the famous Musaeum of Alexandria—and with living quarters for scientists, philosophers, and translators. The latter had the specific task of rendering into Arabic scientific and philosophical texts of the ancient Greek civilization; among these, some foundational mathematical works: Euclid's *Elements*, Apollonius' *Conics*, the complete works of Archimedes, Diophantus' *Arithmetica*, and many others. These translations, observes historian Roero, not only "served as springboard for the continuation of already existing mathematical activities in the Arab world," but they were also "the means that allowed classical Greek works to be known in the West."[35]

One of the most prominent guests of the House of Wisdom was the Persian mathematician and astronomer Muhammad ibn Musa al-Khwarizmi, who lived between 780 and 850. Of his life little is known, apart from his having stayed at the Baghdad academy and the fact—as his name seems to indicate—that he was a native of Khwarizm (today Khiva) a city of the present Uzbekistan. Of his scientific production, on the other hand, some half a dozen texts have survived, among which two—one on arithmetic and the other on algebra—had a profound and permanent impact on the history of mathematics.

Al-Khwarizmi's arithmetic treatise did not come to us in the original Arabic version but in various Latin translations and twelfth- and thirteenth-century paraphrases of his text. On the basis of one of these, a manuscript kept in the Cambridge University Library, in England, the mathematician and historian of science Baldassarre Boncompagni published his own edition of the treatise in 1857, titled *Algoritmi de numero indorum*[36] (Algorithms on the Indian Numerical Calculus). The term "Algoritmi" is none other than the Latinized version of the Persian mathematician's name, which in the Cambridge manuscript appears as "Algorizmi." The document begins in fact with the words: *"Dixit Algorizmi: Laudes deo rectori nostro atque defensori dicamus dignas"* (Thus says al-Khwarizmi: Let us address to God, our Lord and protector, the praise He deserves).[37] Today, *algorithm* refers to a step-by-step procedure to solve a problem, but up until the seventeenth century the word (also in the variant *algorism*) was used to designate the processes of calculation with the numerals of the positional decimal system. This had been invented in India, but al-Khwarizmi gave of it "so full an account that he probably is responsible for the widespread but false impression that our system of numeration is Arabic in origin,"[38] rather than Indian.

Al-Khwarizmi's algebra book, for its part, written between 813 and 833, reached us thanks to a 1342 Arabic manuscript (presently kept in the Bodleian Library in Oxford) and a few earlier Latin translations.[39] The title of the Arabic copy is *Al-kitab al-mukhtasar fi hisab al-jabr wa'l-muqabalah* (The Compendium Book on Calculation by Completion and Balancing). Our word *algebra* comes precisely from the Arabic *al-jabr* in this title. We shall return to its meaning shortly; first we wish to explain why, according to so many experts, this book should be considered as the true "act of birth"[40] of the discipline of algebra, and consequently why the Persian mathematician, rather than Diophantus, "is entitled to be known as the 'father of algebra.'"[41] At first glimpse, it would appear that the opposite is true: al-Khwarizmi's work—which for convenience we will simply call *Algebra*—is entirely rhetorical, with no mention whatsoever of symbols or abbreviations, and therefore might be seen as a regression compared to the syncopated algebra of the Greek author. For the experts, however, the key element in favor of al-Khwarizmi is the fact that in *Algebra* there appears for the first time, limpid and explicit, the notion of an equation as a mathematical entity by itself, separate from any application. Roero puts it in these terms:

> Among the major concepts employed [in *Algebra*] we find the notions of first- and second-degree equation with numerical coefficients, a

novel characteristic with respect to earlier mathematics. It is no longer a question, as in the case of the Egyptians, Babylonians, and Greeks, of solving arithmetic and geometric problems that can be expressed in the form of equations; on the contrary: equations are the starting point and the problems come later.[42]

Algebra is about much more than equations. The treatise, intended by the author to be used for solving problems of everyday life, opens with an introduction to commercial contracts and related calculations; the section on algebra is followed by a chapter on plane and solid geometry and a long section devoted to questions of estate distribution. But the book's most significant achievement is certainly the detailed analysis of first- and second-degree equations made by the Persian mathematician.

Since al-Khwarizmi does not use symbols, he has to resort to specific words to indicate the terms of an equation: our present-day unknown *x* is called *say'* (thing) or *gizr* (root), from the Arabic word for the root of a plant (and also used to designate the square root); the square of the unknown (x^2) is indicated with *mal* (good, or possession), while numbers are called *dirham*, coming perhaps from drachma, the Greek currency unit. This terminology—translated into Latin as *res* or *radix*, *census*, and *numerus*, respectively—will be used with identical meaning by Western mathematicians in the Middle Ages.

In Tartaglia's and his contemporary's vernacular Italian, *cosa* (thing) is still used to designate *x*—thus the expression "rule of the thing" to refer to algebra[43]—*censo* for x^2, and *numero* (number) for the "constant term" of the equation.[44] The word "equation," to indicate an equality involving an unknown to be found, was introduced by Leonardo Fibonacci; it is first used in this sense in his 1202 *Liber abaci*.[45]

Going back to al-Khwarizmi, he examines six types of equation in his *Algebra* (his verbal expressions are listed on the left and the corresponding modern notation on the right, where the coefficients *a*, *b*, and *c* are positive numbers):

Census equal to thing　　　$ax^2 = bx$
Census equal to number　　$ax^2 = c$
Thing equal to number　　　$ax = c$

Census and thing equal to number	$ax^2 + bx = c$
Census and number equal to thing	$ax^2 + c = bx$
Thing and number equal to census	$bx + c = ax^2$

Perhaps a present-day college student would be a little surprised and disconcerted by the multiplicity of cases, knowing that all this apparent variety can be reduced to the single equation $ax^2 + bx + c = 0$. Assuming that the coefficients a, b, and c denote arbitrary numbers, it is the most general form of a quadratic equation (except when $a = 0$, for it then becomes a first-degree equation). Its solution formula can be found in every elementary algebra textbook:

$$x = \frac{-b \pm \sqrt{b^2 - 4ac}}{2a}$$

This is how the universal formula for solving all second-degree equations is presented today. It is actually very simple: for a given quadratic equation, all we need to do to obtain the solutions is replace a, b, and c by their numerical values and carry out the indicated operations. The $\pm$ sign means that two calculations must be performed, one for each sign, leading to two possible solutions; in the first case, the square root is added; in the second, it is subtracted. It is of course understood that the equation must be written in the general form $ax^2 + bx + c = 0$. Al-Khwarizmi never took this general form into consideration because he did not contemplate the possibility for the coefficients of an equation to be zero or negative; the proliferation of equations in his writings is nothing but the direct consequence of his allowing only positive coefficients.

In his *Algebra*, the Persian mathematician reduces every first- and second-degree equation to one of the types listed above, using the two techniques mentioned in the title of his treatise: *al-jabr* and *al-muqabalah*. The first one, *al-jabr* (literally "completion" or "filling," *restauratio* in Latin) is the transposition of a term of the equation from one side of the equal sign to the other, as for example, when going from $x^2 - 5x = 4x$ to $x^2 = 4x + 5x$ (adding $5x$ to each side results in the elimination of $-5x$ on the left). *Al-muqabalah* ("balancing," *oppositio* in Latin) consists in the reduction of similar terms occurring in one or both members of the equation; in the above example, it allows us to go from $x^2 = 4x + 5x$ to $x^2 = 9x$.

Regarding the terms *al-jabr* and *al-muqabalah*, historian Roero writes:

> The terms *al-jabr*, from which the word "algebra" was derived, and *al-muqabalah* appear in the titles of subsequent works on algebra by Islamic mathematicians as a general description of texts on the theory of equations. Their use will extend to the West beginning with Fibonacci and until the end of the 15th century, first in abbaco schools' textbooks and later in printed volumes, always in the sense of "theory of equations." The use of the term "algebra", with a broader meaning, will continue until today, while the word *al-muqabalah* will no longer appear after the 15th century.[46]

For each of the six types of equation listed above, al-Kwharizmi describes, in the form of verbal recipes, the corresponding solution rules—all of them equivalent to the universal formula[47]—accompanied by a geometric demonstration. Let us listen again to Roero:

> It is true that methods to solve first- and second-degree equations were known before the Arabs, but it was never felt necessary to develop a theory of equations; these were only considered in connection with the specific problem from which they arose. Al-Khwarizmi, on the other hand, studied equations as a mathematical object in itself, set up a classification, and gave a solution procedure for each case.[48]

Since he did not allow null or negative numerical coefficients, al-Khwarizmi did not accept null or negative solutions either. And yet, he was familiar with the concepts of zero and negative number, introduced and developed in India in the seventh century CE, notably by the mathematician and astronomer Brahmagupta. To this eminent scientist we owe gaining the "complete awareness of the existence of zero as a real and proper number."[49] "The systematized arithmetic of negative numbers [to represent debts in financial calculations] is first found in his work."[50] Why al-Khwarizmi persisted in ignoring null and negative solutions is not clear; it is possible that he considered only positive solutions because they are the only ones to admit a geometrical interpretation. Besides, Roero adds: "... this standpoint will remain unchanged also among the algebraists of the late Middle Ages and the Renaissance, and will not be called into question until the seventeenth century. In René Descartes'[51] *Géométrie* (1637) we still find examples of quadratic equations of which only the positive solution is given."[52]

Al-Khwarizmi's detailed treatment of first- and second-degree equations did not carry over to those of third degree. These were studied with great interest by other Islamic scholars of the Middle Ages, among whom stands out the Persian mathematician, astronomer, philosopher, and poet Omar Khayyam.

Originally from the city of Nishapur, in present-day northeastern Iran, Khayyam—who lived between 1048 and 1131—is known in the West mostly for his poems in quatrains. He was also the author of the first general theory of third-degree equations, similar to that of al-Khwarizmi for those of lower degree, which he presented in his work *Risalah fi'l-barahin ala mas'il al-jabr wa'l-muqabalah* (Treatise on Demonstration of Problems of Algebra),[53] written in around 1074. He set up a classification of cubic equations in fourteen different types (where x, x^2, and numbers were indicated with the terms already mentioned, and x^3 with the word *ka'b*, cube).[54]

Cube equal to number	$x^3 = c$
Cube and thing equal to number	$x^3 + bx = c$
Cube and number equal to thing	$x^3 + c = bx$
Thing and number equal to cube	$bx + c = x^3$
Cube and census equal to number	$x^3 + ax^2 = c$
Cube and number equal to census	$x^3 + c = ax^2$
Census and number equal to cube	$ax^2 + c = x^3$
Cube and census and thing equal to number	$x^3 + ax^2 + bx = c$
Cube and census and number equal to thing	$x^3 + ax^2 + c = bx$
Cube and thing and number equal to census	$x^3 + bx + c = ax^2$
Census and thing and number equal to cube	$ax^2 + bx + c = x^3$

Cube and census equal to thing and number $x^3 + ax^2 = bx + c$
Cube and thing equal to census and number $x^3 + bx = ax^2 + c$
Cube and number equal to census and thing $x^3 + c = ax^2 + bx$

Just as al-Khwarizmi had done, Omar Khayyam imposed the condition that the numerical coefficients be positive, and he equally considered only positive solutions; this is again the reason for the large number of cases, which today are reduced to a single one: the general third-degree equation $ax^3 + bx^2 + cx + d = 0$, with arbitrary numbers a, b, c, and d (where a must be different from zero to prevent the equation from becoming one of lower degree).

With exceptional skill and sophisticated arguments, Khayyam was able to solve *geometrically* all equations of the types listed above using the so-called conic sections (circle, ellipse, parabola, and hyperbola). For example, he showed that from a geometric point of view, the solution of the equation $x^3 + bx = c$ is the length of the segment AB in the figure below, obtained from the intersection of a parabola and a circle suitably constructed from the values of the coefficients b and c.

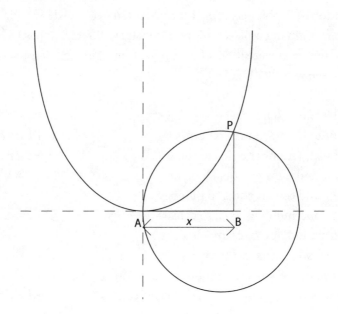

For all their ingenuity, and as worthy of admiration as they are, Khayyam's remain only geometric solutions to cubic equations, where the unknown is graphically visualized on the basis of properties of certain curves. The Persian mathematician himself frankly admits his failure to attain the more ambitious goal he sought: to obtain numerical solutions for third-degree equations by means of a general algebraic formula; a universal formula like the one for quadratic equations, where the unknown of any cubic equation could be found simply by numerical substitution of the coefficients. Such was the real challenge. Had he succeeded and discovered the formula for cubic equations, it would have represented the first truly new result, the first breakthrough in algebra since the time Babylonians learned how to solve quadratic equations.

After Omar Khayyam, plenty of mathematicians tried their hand at third-degree equations, but for nearly four centuries no one got beyond solving particular cases, geometrically determining the unknown, or calculating approximate numerical solutions. In his 1225 work *Flos* (Flower), Leonardo Fibonacci tackled the complete cubic equation $x^3 + 2x^2 + 10x = 20$ and gave an approximate solution, exact to the tenth decimal place ($x = 1.3688081078$), but with no indication of how he obtained it.[55]

And so, on the threshold of the sixteenth century the formula for cubic equations remained an impenetrable mystery, and the task to discover it appeared so formidable as to be considered next to impossible by the eminent Luca Pacioli. Omar Khayyam proved a better prophet, for he wrote: "Maybe one of those who will come after us will succeed in finding it."[56]

CHAPTER 3

The Venetian Challenge

After almost two decades in Verona, Niccolo Tartaglia moved to Venice in 1534, "into the new houses of San Salvatore."[1] The reasons for his relocation to the capital of the Most Serene Republic are obscure, but we may assume it was motivated largely by his desire to have his works printed. Indeed, his entire scientific production would be published in Venice, Europe's major typographical center of the time.[2]

Introduced in Venice by the German engraver Johann von Speyer in 1469, the art of printing underwent a rapid and impressive development. At the beginning of the sixteenth century the city was home to more than 200 typographers, while Milan, for instance, counted only sixty, and both Rome and Bologna a mere forty each.[3] Besides, conditions were particularly favorable for the book industry and commerce as a consequence of Venice's geographic location, merchant activities, widespread economic affluence, and the lively intellectual ferment characteristic of the city of the Doges. Moreover, notes historian Carlo Maccagni, at the time "Venice had its own cultural life,"[4] independent from the University of Padua—the only university in the Venetian Republic—and based on two didactical structures: the School of San Marco, devoted essentially to humanistic studies, and the School of Rialto, specializing in science and philosophy. In addition to these two schools, in October 1530 the Senate of the Republic created a public mathematics course, justifying its decision in the following terms: "In every epoch, and in every nation, the study of the *bonnes lettres* [i.e., the knowledge of respected Greco-Latin authors] was and still is manifest; it is thanks to them that all good art is acquired, that the youth learn the proper customs and men become useful to the administration of the State [. . .] Among the liberal arts,[5] that which above all others must be sought for being the most certain and of the greatest value to the life of humans, is called mathematics."[6]

Giovan Battista Memo, known for having been the first to translate into Latin Apollonius' *Conics*,[7] was appointed to the new chair of mathematics. It seems entirely plausible that Memo—who died in 1536—was succeeded by Tartaglia; although no official document was ever found to confirm it, a number of passages in the writings of the Brescian mathematician point in that direction. For instance, we know for certain that in August of the same year (1536) Niccolo taught the geometry of Euclid at the Venetian Church of Saints John and Paul, and three years later lectured on statics and ballistics.[8]

Tartaglia's scientific activities in Venice were plentiful and varied: alternating between public and private teaching; attending to his own research; fulfilling multiple consulting engagements; and responding to numerous requests to discuss mathematical, mechanical, topographic, and military problems. In fact, Niccolo had already been brought into the limelight shortly after his arrival in Venice when Antonio Maria Fior, a highly respected scholar in Venetian society, challenged him to a public algebraic duel. The contest received intense attention in Italian mathematical circles, as much for its content as for its sensational outcome.

Not much is known of Antonio Maria Fior's life. Born in Venice during the last quarter of the fifteenth century, he was the son of a certain "Master Pelegrino."[9] We know that early in the following century, Fior studied arithmetic and geometry at the University of Bologna, seat of the "most famous school of mathematics in Renaissance Italy,"[10] and that by 1520, he had earned a reputation as a remarkable master of "abbaco and *quaderno* (a sort of bookkeeping)."[11] It was "especially his skill for teaching 'quaderno', that is, the compilation of registers for the economic administration,"[12] that brought him into public attention in connection with the then new discipline of accountancy; his services were sought mostly for his expertise in this domain. In 1516, the young Matthäus Schwarz, an accountant for the Fuggers, a powerful family of German bankers, was on a study trip in Italy. Not satisfied with the lessons he took in Milan and Genoa, he traveled to Venice to learn from Fior how to "record the entries."[13] Schwarz would later write a report of his profitable experience with Fior, mentioning, however, the low overall cultural level of his Italian teacher.[14]

At the beginning of 1535, Antonio Maria Fior began a dispute with Niccolo Tartaglia, challenging him to a mathematical duel to be conducted according to the rules of the time. He proposed thirty problems to Tartaglia, who in response sent the challenger thirty other questions. As was customary, the winner would be whoever answered the most questions in a given period of time. On February 22, 1535, Fior and Tartaglia entrusted their respective lists of questions to Iacomo Zambelli, a notary public in Venice, agreeing to hand over the solutions forty or fifty days later.[15] The stakes of the contest? First of all, honor and reputation, and then a lavish dinner at the tavern for each unsolved problem, with the bill to be footed by the unfortunate contestant who capitulated before his rival's question.

In the ensuing months, echoes of the Venetian challenge spread well beyond the city, carrying the news of its resounding outcome: Tartaglia had literally humiliated Fior by solving in a couple of hours all thirty problems posed by his opponent, while the latter had not been able to answer a single one of Niccolo's questions (the victor, the story goes, magnanimously gave up his right to the dinners). In fact, far more astonishing was the topic on which Fior had challenged Tartaglia. Such details about the contest slowly trickled out of Venice and eventually reached Tartaglia's birthplace, Brescia, and the indiscreet ears of one of Niccolo's old acquaintances: the imperishable Messer Zuanne de Tonini da Coi.

After the controversial epistolary exchange of 1530, the relationship between the two Brescians must have improved a little, given the fact that at the end of 1536 Tonini da Coi travelled to Venice and was received by Tartaglia, who recounted to him in person the particulars of his triumphal victory.

In the aforementioned first edition of *Quesiti* (1546) there is an account of the dialogue between Niccolo and Zuanne during their meeting in Venice on December 10, 1536. Zuanne begins by reporting what he had heard and understood about Fior's challenge, and admitting his incredulity regarding the conclusion of the duel:

> I heard many days ago that you had entered into a challenge with Maestro Antonio Maria Fior, and that finally you had agreed on this: that he would propose to you thirty really different problems written under seal and in the custody of Maestro Iacomo di Zambelli, notary; and that, similarly, you would propose to him thirty other, truly different ones. And so you did, and each of you was given forty or fifty days to solve the said problems; and you established that, at the end of this period, whoever had solved the largest number of problems would receive the honors, in addition to whatever small sum you had waged for each problem. And it has been reported and confirmed to me in Brescia that you solved all his thirty problems in two hours, which is something I find hard to believe.[16]

"Everything you have been told is true,"[17] replied Tartaglia, who right away explained how and why he had been able to perform a feat that was to be the peak of his scientific career and a decisive turning point in the history of algebra. Niccolo then revealed the nature of the problems proposed by Fior and which he had so brilliantly solved, leaving Messer Zuanne even more astonished and perplexed:

> The reason I could solve his thirty problems so rapidly is that all of them lead by algebra to thing and cube equal to number [i.e., all the problems could be reduced to a cubic equation of the form $x^3 + bx = c$]. He believed I couldn't solve any of them, because Fra Luca [Pacioli] deemed such chapters[18] impossible to solve with a general rule; and I, by a stroke of luck, only eight days before the date to put the thirty and thirty problems under seal with the notary, had found the general rule for such chapter.[19] This took place last year,

that is, on 12 February 1535 (although in Venice the year was 1534).[20] Such discovery being fresh in my mind, it was ready and familiar, and that is why I could solve all thirty problems with so much celerity, or swiftness.[21]

According to this account, Niccolo Tartaglia discovered the formula for the solution of third-degree equations of the type $x^3 + bx = c$ on February 12, 1535, a formula that a few years later would prove crucial for solving *any* cubic equation. It was an epoch-making, extraordinary result: the first true algebraic discovery since the time of the Babylonians, one that would open up new and boundless horizons for algebra and that posterity would remember as the most significant mathematical result of the sixteenth century.

A word of caution, however: Renaissance mathematical duels obeyed a gentlemen's agreement: a contender could not propose to his opponent problems he did not know how to solve himself. Therefore, if Fior felt confident in challenging Tartaglia with problems involving cubic equations of the type mentioned earlier, then the Venetian abbaco master must have discovered the fateful formula before his Brescian rival—unless Fior had not played fair and sought to intimidate Tartaglia with problems beyond his own ability to solve, just as Tonino da Coi had done five years earlier. On the other hand, if we rule out this possibility, how could a teacher known only for being a good expert in accounting calculations—and, what is more, not particularly cultivated or at ease with algebraic abstractions—been capable of such an extraordinary undertaking?

Faced with Messer Zuanne's increasing curiosity, Tartaglia gave out some clues that clarify this point. But in fact his explanation raised new and rather explosive questions:

He [Fior] was boasting of having found such a rule to scare me. At first, I did not believe him, because he did not possess the science but only great experience; and by experience alone I did not deem him capable of finding such a rule on his own. But, to make me believe that I should fear him even if he didn't possess the theory, he claimed that thirty years ago a great mathematician had revealed to him the secret. I had a feeling that it might be true, and for this reason I put off my studies, care, and work to find the rule for that chapter [i.e., for that type of equation].[22]

Tartaglia's opinion of Fior's intellectual capabilities—he did not possess the science but only great experience—is consistent with that of Matthäus Schwarz, the German accountant who had taken lessons from Fior. Once established that Fior lacked the education and fertile mind required to find the formula on his own, and assuming that he already knew it when he threw down the gauntlet to challenge Tartaglia, under what circumstances did he acquire this knowledge? And, especially, who was the "great mathematician" who had revealed to him the secret thirty years earlier, and who therefore should have been credited with the exceptional discovery? How was it possible for such a breakthrough to remain secret through all this time? Was Fior the only one to benefit from the confidences of the mysterious personage? And why, after his initial hesitation, had Tartaglia taken Fior's surprising revelation seriously, causing him to put off his "studies, care, and work" to find—or find again—the crucial formula?

Their dialogue goes on without throwing any light on these puzzling questions, but it revealed some astonishing news: the day after having discovered the solution formula for third-degree equations of the type "thing and cube equal to number," that is, on February 13, 1535, Tartaglia had done the same for the case "thing and number equal to cube" ($bx + c = x^3$, in today's notation). It was as if in those first decades of the sixteenth century, the wall of cubic equations, which had remained impenetrable for three thousand years, was suddenly beginning to crumble.

MESSER ZUANNE: You were lucky to have found the solution so suddenly, because if you hadn't, you would have been condemned by the ignorant masses, but not by intelligent people. Because discovering a single secret does not make a man wise; since science is the knowledge of general truths, not particular ones; because these are infinite in number, and so it's impossible to know every one of them. But tell me, what was the topic of the thirty problems you proposed to him?

NICCOLO: They were all entirely different. I did so to show him
that my knowledge is universal, and that I am versed in more
than one, two, or three particular inventions or secrets, and that
on these I could have proposed to him ten thousand problems,
not just thirty. By choosing them (as I said) all truly different
from each other, I wanted him to know that I did not esteem or
fear him in any way.

MESSER ZUANNE: And how many of your problems did he solve?

NICCOLO: He couldn't find the answer to any of them. It's true that
he was boasting that he had solved them, but he never allowed
me to see his solutions. To cover up the matter, he asked me to
choose some of his friends to judge whether he had really solved
them or not. All of them concluded that he had lost the contest,
so I publicly made him a gift of the prize at stake.[23]

At this point, Tonini da Coi cannot hold back any longer and asks
Tartaglia for something he had impatiently wished for since he heard
the news of the Venetian challenge.

MESSER ZUANNE: Please, give me in writing the thirty problems
he proposed to you, with their solutions, and the thirty problems
you proposed to him.

NICCOLO: If I had the time to copy them down, I would be happy
to give you his problems, but not my solutions, because if you
saw them, you would immediately understand the rule. If you
still wish only the problems, go to see the notary, offer him a
little something and he will oblige. As for the thirty questions
I asked him, I don't have them any more, because I took them
right away to the notary to be kept under seal and did not make
any copies. I wouldn't be able to tell you half of them without
asking the notary for a copy.[24]

Tartaglia was therefore not willing to reveal the solution process he
had used to answer Fior's questions for fear that Messer Zuanne would
infer from it the solution formula for third-degree equations of the form
$x^3 + bx = c$, which for the moment Niccolo had no intention to give

away. As mentioned before, in those days it was the custom among mathematicians to refrain from announcing their results; keeping them secret could help attract students as well as serve as a weapon in public contests. Tartaglia followed the custom, as obviously did the unknown "great mathematician" mentioned by Fior, if he existed at all.

Reluctant as he was to share his discovery with Tonini da Coi, in January 1539 Tartaglia would have a different attitude toward someone we shall meet later—and to whom he gave the list of Fior's thirty problems. Here's a small sample of these (the resulting equation appears in square brackets):

1. Find me a number that added to its cubic root make 6.*
   $[x^3 + x = 6]$
2. Find me two numbers in double proportion [i.e., such that one number is twice the other] such that the square of the largest number multiplied by the smallest and added to the first two numbers is 40.† $[4x^3 + 3x = 40]$
6. Two men have earned 100 ducats and must share this sum as follows: one must receive the cubic root of the other. I ask how much each of them gets of the sum they earned. $[x^3 + x = 100]$
13. A man[25] lends a certain amount of money on this condition: that after one year the interest due be the cubic root of the principal; at the end of the year the lender receives 800 ducats in capital plus interest. I ask the amount of the loan.
    $[x^3 + x = 800]$
20. The sum of the areas of two squares equals 26, and the smallest area is the cubic root of the largest one. I ask for the area of the largest square. $[x^3 + x = 26]$[26]

Not only the foregoing examples, but all of the questions Fior asked Tartaglia were algebraically reducible to a cubic equation of the form $x^3 + bx = c$.[27] Seen in this light, it was not so surprising that Niccolo was

---

* Denoting by x^3 the required number and by x its cubic root results in the equation whose solution is the answer to the problem. The same applies to the other problems.

† The relevant equation is found by denoting the two numbers by x and $2x$ and applying the usual rules of algebraic calculus.

able to answer them in such a short time, having discovered the general formula for that type of equation. The fact that Fior had bet all his chips on this single equation, along with the accompanying risks, is an indication not only of his presumptuous conviction of check-mating Tartaglia in a single move but, presumably, that he had only one arrow in his quiver. Surely, it was a sharp one; but it must not have been easy for the unfortunate Venetian accountant to digest such a disastrous and unexpected defeat.

"Truly different from each other" were instead the questions Tartaglia posed to Fior, four of which Niccolo finally revealed to Messer Zuanne:

1. Find me an irrational quantity that multiplied by its [square] root plus 40 makes a rational and discrete number [that is, an integer].*
2. Find me an irrational quantity that multiplied by 30, minus the [square] root of the said quantity, makes a rational and discrete number.
3. Find me a quantity that added to the quadruple of its cubic root makes 13.
4. Find me a quantity such that after subtracting 3 from its cubic root there remains 10.[28]

These problems lead respectively to the following equations (where n denotes a positive number):

1. $x^3 + 40x^2 = n$.
2. $x^3 + n = 30 \, x^2$.
3. $x^3 + 4x = 13$.
4. $x^3 = 3x + 10$.

A peculiar fact emerges when we consider the third equation: it has the same form as those proposed by Fior to Tartaglia, a clear sign that, after having discovered the solution formula, Niccolo wished to test

---

* The equations for solving this and the next problem are obtained by denoting by x^2 the quantity to be found and by x its square root.

whether his rival really knew it and could solve at least that equation. Otherwise, not many other scenarios can be envisaged: either Fior was given the formula but he did not understand it and therefore could not apply it, or else the story of the "great mathematician" from whom he presumably learned the secret was a pure invention.

A second striking element concerns the first two equations above: Tonini da Coi realizes—with understandable astonishment—that Tartaglia had also managed to obtain the solution formula for cubic equations of the form "cube and cenno equal to number" $[x^3 + ax^2 = c]$ and "cube and number equal to cenno" $[x^3 + c = ax^2]$. "This I found at the end of 1530, when I was living in Verona,"[29] declared Niccolo, reminding Messer Zuanne—who at first pretended not to remember— that it was precisely one the problems he had sent him that led to the result: "You asked me to find a number that multiplied by its [square] root plus 3 makes 5, which leads, as you know, to one cube plus 3 cenno equal to 5 $[x^3 + 3x^2 = 5]$."[30] Tonini da Coi having finally admitted to the facts—"I remember"[31]—Tartaglia confirmed that it was Coi's problem that triggered his discovery of the general solution for the equation $x^3 + ax^2 = c$, and soon afterwards of that for equations of the form $x^3 + c = ax^2$. Niccolo, though, was being a little too smart. . . . Let us see why.

Tartaglia and Tonini da Coi got together once again in Venice on December 15, 1536, five days after their previous meeting. Messer Zuanne wasted no time in insisting: "Messer Niccolo, I shall go back to Brescia in a few days, but before I leave, please, would you let me have one of your four problems and its solution?"[32] And Tartaglia's blunt reply: "You should know, Messer Zuanne, that inventions are difficult, and developing them is easy. Therefore, having strived to find those particular facts, it does not appear to me that I should so easily make them public, especially if I could not reap any honor or benefit."[33]

Nonetheless, Niccolo went on to assure him that he had no intention of hiding his discoveries but was waiting only to complete some other work before making them available: "It is most certainly true that keeping these inventions totally buried is not fair. Rest assured that it is not my intention to do so, but to publish them for everyone to see. As soon as I finish some work already in progress, I hope to make good on my promise."[34]

Tartaglia then makes him a provocative proposition: to reveal one of his secret formulas for each question Zuanne posed to him that he [Tartaglia] would be unable to answer:

> So that you wouldn't think that I value my inventions more than I should, I offer you, each time I will not be able to solve one or several of your problems, to exchange one solution of yours against one of mine. This is not a small offer: swapping something general (from which you can not only form infinitely many cases, but also discover the rule to solve many others) for a particular case.[35]

Perhaps not completely aware of having been mocked by Niccolo (given the obvious difference in mathematical skills), the intrepid Messer Zuanne did not hesitate: "Your proposition is an honest one, and therefore I wish to propose two delightful problems; if you don't know how to solve them, I will show you the answer and you will show me the rule for those chapters [i.e., equations] of yours, especially the one for thing and cube equal to number $[x^3 + bx = c]$."[36]

Eager to catch the coveted prey, Zuanne stated without hesitation two geometry problems about triangles and circles, convinced of having outwitted Tartaglia: "Now, if you show me your rule for cenno and cube equal to number $[x^3 + ax^2 = c]$, I will show you how to solve these two problems; they really are lovely, and difficult."[37]

Tonini da Coi's elation was short-lived: "These are easy problems, because if you give me one hour I shall solve them for you,"[38] replied Tartaglia, and he went on to inflict a further blow to his rival.

Niccolo dug up an incident that had taken place the previous year, more precisely on 12 September 1535, when a certain Dominico da Underzo forwarded him some questions asked "anonymously" by "someone in Brescia,"[39] whom Tartaglia immediately identified as the proverbial Messer Zuanne. Among the problems he had sent there was one reducible to a fourth-degree equation: $x^4 + 8x^2 + 64 = 160x$. Tartaglia knew perfectly well that Zuanne was incapable of solving "with a general rule" a fourth-degree equation. And so, one year later, Niccolo seized the opportunity to reprimand him, using the blade of sarcasm, for his incurable and pernicious habit of challenging others with problems he could not solve himself:

> If you show me how to solve this problem[40] with a general rule, I shall explain to you the method and the general rule for solving whichever equation you prefer from the four I showed you before.[41] Although I know you will not accept this offer, because you don't know how to solve the problem. It is remarkable that you are unable to get rid of your old habit.[42]

Tonini da Coi tried awkwardly to defend himself; then, rebuked once again, stubbornly resumed his plea:

> MESSER ZUANNE: And I tell you that it is not less beautiful to know how to demonstrate the impossibility of an irresoluble case than to solve a resolvable one.
>
> NICCOLO: Your excuse is not a good one, because you did not propose it as being impossible,[43] but as something you did not understand or know how to solve, and even less know how to prove it is impossible. But neither you nor anybody else could convince me by a reasoned argument that it is impossible to solve. What I mean is that not only do I believe that your problem is possible, but that after attending to certain matters of mine I expect to find the general rule, as I did for those you sent me to Verona.
>
> MESSER ZUANNE: You would have achieved a great deal if you found it. Now, because I must be on my way shortly, if you would be so kind as to let me have at least one of those four problems of yours solved, I promise that as soon as I get to Brescia I shall send you some lovely problems; if you do not know how to solve them, I shall also send you the solutions. And if you happened to have other questions besides those regarding your new chapters [i.e., equations], please give them to me, and if I know how to answer them I shall send you back the solutions in writing.[44]

Somewhat unexpectedly, this time Tartaglia gives in to the persistent and repeated demands of Messer Zuanne—"your plea compels me to give you a little satisfaction"[45]—and decides to reveal the answer to one of the questions he addressed to Fior: "Find me an irrational quantity that multiplied by its [square] root plus 40 makes a rational and discrete number,"[46] which in modern notation becomes $x^3 + 40x^2 = n$, where n denotes some positive integer (a "rational and discrete" number). In particular, Niccolo showed Zuanne the solution of the equation for $n = 2888$. He justified the choice of this value by the requirement that the unknown quantity should be an irrational number, which indeed it is: the solution, as Tartaglia told da Coi, is $x = \sqrt{77} - 1$.

It is important to emphasize that Tartaglia disclosed only the result, without offering any explanation as to the steps that led to it. In fact, he had earlier pointed out to Tonini da Coi that knowing the answer would allow him to infer the solution method, which Zuanne did succeed in deducing—and which he communicated to Tartaglia in a letter of January 8, 1537—earning Niccolo's tacit approval.[47]

Based on an analysis of the solution of the equation $x^3 + 40x^2 = 2888$ provided by Tartaglia and in the procedure figured out by Tonini da Coi to obtain it, mathematics historians have unanimously concluded that, contrary to his claim, in 1530 Tartaglia had *not* found any general rule for cubic equations of the form $x^3 + ax^2 = c$ (cube and cenno equal to number). He had developed a technique only for generating equations of that form that admit a particular irrational solution chosen in advance.[48]

Niccolo's claim of possessing the general solution formula for such equations may be dismissed as "innocent boasting"[49] on his part, but it can also be interpreted as a calculated, expedient trick to gain the upper hand in mathematical contests, as in his duel with Fior.[50] Let us be clear: the contestant who challenged his rival with a given problem was not necessarily expected to know the general rule for solving all others of the same type; it was enough for him to think up a question based on a solution chosen in advance and not easily recognizable. In that case, it was up to his opponent to find the general solution rule in order to avoid defeat. Seen in this light, Tartaglia's was a truly cunning scheme.

Before moving forward with the story, let us try to put some order among the facts reported so far. Tartaglia began to focus his interest on cubic equations in 1530, in Verona, prompted by two problems that Zuanne de Tonini da Coi had addressed to him. On that occasion, Niccolo developed a method to construct third-degree equations of the form $x^3 + ax^2 = c$ and $x^3 + c = ax^2$ from given irrational solutions. But he did not go beyond this: despite his own claims, he did not discover their general solution formulas. Five years later, in Venice, he was challenged to a mathematical duel by Antonio Maria Fior, who pretended to know the rule for solving cubics of the form $x^3 + bx = c$, a rule—so he claimed—secretly revealed to him by a mysterious "great mathematician." Allowing for the possibility that Fior was telling the truth, Tartaglia accepted the challenge and was confronted with thirty problems leading to equations of the above form. He found the general formula, this time for real; otherwise he would not have been able to solve them, as he did, in almost no time at all and win the contest.

Once he got back to Brescia, in his January 8, 1537, letter Tonini de Coi resumed pestering Tartaglia with requests, not satisfied with all that Niccolo had revealed to him in Venice.

His most fervent desire remained the same: the formula for equations of the form $x^3 + bx = c$. To achieve his end, the unflappable Messer Zuanne resorted to a well-proven tactic, flattery, adopting a tone of excessive praise. After beginning the letter with a "Dearest Messer Niccolo"[51] and before closing it with a warm "your good-hearted brother,"[52] he wishes to express his gratitude for "so much benevolence" and the "friendly politeness"[53] he received in Venice from Tartaglia, whom he had found to be a man "full of kindness, of loyalty, of courtesy, and finally of honest and noble conversation."[54] He added that he could have asked Fior for the formula (it seems almost certain that he actually did, to no avail),[55] but he had preferred to first turn to him, to "Master Niccolo," because "I value your friendship much more than his on account of your superior virtuousness"[56] and also because we both "come from the same town"[57] (i.e., from Brescia).

On the other hand, continued Messer Zuanne, this time with clear and pointed malice, Tartaglia had no reason to protect that formula so jealously, given that it wasn't really his "own invention"[58]: someone had discovered it first, and the glory belonged to him ("Are you not aware that only first inventions are praised and awarded prizes by those who know?"[59]). Why, then, persist in keeping a secret already known to others?

Tartaglia did not reply to Tonini da Coi's letter. Far from capitulating, in successive communications Messer Zuanne continued to send Niccolo problem upon problem, besides, of course, begging him to reveal the famous formula. Eventually, Niccolo decided to reply, and in a letter dated March 3, 1537, begins by letting his impertinent correspondent know that to satisfy his every request the writing would have filled "a ream of paper," which my "daily and nightly occupations do not allow me," having better things to do than "spend my days solving your problems with no benefit or honor whatsoever."[60] Having in this way made himself clear, Tartaglia responded to Messer Zuanne's remark on the paternity of the formula for $x^3 + bx = c$:

> You say (for the purpose of slandering my invention) that only first inventions are praised by those who know, and that mine is not a proper one since someone knew it before me.
>
> To which I reply that you are neither certain nor have any evidence that my rival possessed such secret, except for the fact that his thirty problems all led to that difficult case. This is not proof that he possessed or knew the secret; because many, to confound his opponent, are in the habit of asking questions they neither understand nor know how to answer, just as you did when I was in Verona with your two requests.
>
> But even assuming that my rival knew the formula before me, since I discovered it again with no help from anyone, it may be considered my own invention.[61]

Tartaglia was by now certain that Tonini da Coi's expressions of friendship were not sincere or disinterested, and that all he wanted was to extract from him the secret formula for cubic equations so he could use it to brag to other mathematicians. In this regard, he had received new particulars ("after you left Venice [. . .], I heard of so many of your bravados that it would be too long to recount").[62] As for the multitude of problems Messer Zuanne sent him, Niccolo wrote: "Your incessant questions are nothing but an attempt to discover my weaknesses, and for this reason I decided not to answer any of them until you came in person to Venice [. . .]."[63]

After the March 1537 letter, the relationship between Niccolo Tartaglia and Zuanne de Tonini da Coi was definitely over.

At the end of that same year Tartaglia published his first book, a small volume titled *Nova scientia*[64] that would be remembered as the first mathematical treatment of the motion of projectiles and the first theoretical analysis of ballistics, an art that "hitherto was only practical."[65] As he put it in the preface, Niccolo's interest in the discipline was sparked by a question from a bomber friend, who in 1531 had asked him "how to aim a piece of artillery the farthest it can fire."[66] Without ever having fired "a piece of artillery, an arquebus, a bombard, or a blunderbuss"[67]—Tartaglia despised firearms and military matters, perhaps as a consequence of the dramatic events he went through in his youth—he set out to study the question and after having "properly chewed and ruminated" on it,[68] solved the problem: the projectile will reach the farthest when it is aimed at an angle of 45 degrees above the horizon. This is *Nova scientia*'s most significant result. It is a correct proposition, despite the fact that it was obtained through an erroneous argument. Actually, all the arguments in the book—-which was a resounding success—are completely fallacious from the perspective of present knowledge. This in no way detracts from the fact that the book constitutes "a fundamental stage of the passage from technical to scientific knowledge,"[69] as historian Roberto Maiocchi observes.

Going back to mathematics, as months went by, news about the Fior–Tartaglia challenge reached beyond the borders of the Republic of Venice. On January 2, 1539 Niccolo received the visit of a certain "Zuanantonio, a bookseller,"[70] and messenger on behalf of a scholar from Milan who eagerly wished to know the secret formula for cubic equations. The scholar from Milan was no ordinary scholar. History would immortalize him as one of the most important and controversial figures of the Renaissance: a man with an eccentric personality, who could be variously depicted as a physician, mathematician, natural scientist, humanist, philosopher, astronomer, astrologer, musicologist, magician, dramaturge, dream interpreter, and inveterate gambler: Gerolamo Cardano.

CHAPTER 4

An Invitation to Milan

He was consulted many times by Leonardo da Vinci on questions of geometry. And yet, Messer Fazio was not a professional mathematician. He was a lawyer in Milan, but his multiple interests led him to be active in many fields; besides law and mathematics, these included medicine and occult sciences. Fazio's back was round and he stammered (he too!), and in particular he had a good knowledge of Euclid's works in geometry, which for a while he taught at the University of Pavia. His family came from the town of Cardano—presently Cardano al Campo, some thirty kilometers from Milan—and even if it did not possess large financial resources, it claimed ancient and noble origins.

Fazio was still a bachelor when at the age of fifty-six he met Chiara Micheri, a thirty-six-year-old widow and mother of three sons, with whom he began a romantic relationship. Pretty soon, Chiara—who was living with the family of her departed husband—became pregnant. To prevent the inevitable scandal, Fazio invited her to move to Pavia into the house of his friend, the aristocrat Isidoro Resta, who would pretend to have hired her as a governess. And so, at the home of the Pavian patrician, on September 24, 1501, the woman gave birth to a boy, Gerolamo, after several vain attempts to abort her pregnancy. The baby was born unconscious, but he was "reanimated" with a "bath of warm wine that might have proven fatal to another child."[1]

Shortly afterwards, Chiara received the tragic news from Milan: her three legitimate children had been killed within a short time by the plague that in those years was ravaging Lombardy. Even the wet nurse breastfeeding Gerolamo contracted the disease and died one day later, but luckily the child was not affected. He was disinfected with a bath, this time of "warm vinegar,"[2] and sent to the village of Moirago, not far from Pavia, where he was put in the care of other nurses until the age of three. In Milan, Fazio rented an apartment where he installed Chiara, her sister Margherita, and Girolamo, and that he at long last moved into when his son was seven.

Gerolamo's childhood was anything but quiet: "I was beaten by my father and mother for no reason," will remember the scientist in his autobiography,[3] "and I was sick many times with life-threatening diseases."[4] Fazio and Chiara stopped mistreating their son the moment they decided to live together, but for Girolamo it also meant the beginning of a heavy burden: to follow his father along during his business trips carrying bags, books, papers, and instruments "like a servant."[5] In the meantime, Fazio taught the youngster how to read and write, and he later introduced him to some basic notions of mathematics and other disciplines:

> It was my father who in my childhood taught me the rudiments of arithmetic, and later, when I was about nine, certain quasi-occult notions that he got God knows where, and he also explained to me the astrology of the Arabs. When I was twelve he taught me the first six books of Euclid's *Elements*, but only those notions I couldn't learn by myself. This is all I've learned without going to school and with no knowledge of Latin.[6]

Chiara, for her part, was responsible for the musical education of her son, who would always continue to cultivate music with pleasure and devotion.

When he reached the age of nineteen, the young Cardano entered the University of Pavia to study medicine, although Fazio would have preferred for him a career in law. Gerolamo soon demonstrated exceptional intellectual qualities and rhetorical ability, "holding and winning public debates against his own teachers."[7]

Recalling in his mature years some of the places where his scientific activities took him, he wrote:

> Wherever I happened to be, in Milan, Pavia, Bologna, France, or Germany, since I was twenty three I never met anyone who could measure up to me in a discussion or debate, but I'm not boasting about that. If I had been a stone, the result would have probably been the same, because this is not a privilege of my nature or of its greatness, but it's due to the ignorance of those who challenge me.[8]

At university, Cardano was a serious student and a successful debater. His talent did not go unnoticed and he was invited to lecture on the geometry of Euclid, dialectics, and metaphysics. In 1523, the Franco-Spanish war over the Duchy of Milan[9] resulted in the temporary closing of the university, prompting Gerolamo to move to Venice at the beginning of the following year to pursue his studies at the University of Padua. Here too he immediately stood out for his remarkable qualities and his vast and diversified knowledge, and in the summer of 1524 obtained a baccalaureate degree "in artibus," that is, in medicine and philosophy. Back in Milan for the holidays, he found his father living his final days. The old man rejoiced in the success his son had already achieved in his studies. Fazio passed away at the end of August, not long after having regularized his union with Gerolamo's mother Chiara by marrying her.

In 1525, now twenty-four, Cardano was elected "dean" of students at the University of Padua. The one-year appointment carried certain privileges, but especially various responsibilities, including performing administrative and judicial duties and maintaining regular contact with the academic and municipal authorities. Since the task involved important monetary costs, deans usually belonged to well-to-do families; this was not precisely Gerolamo's case, and he soon regretted having applied for the position. Fazio had left him a meager inheritance, not enough to cover all the expenses he now faced, even added to the money his mother was sending him. To make ends meet, Gerolamo resorted to an old and uncontrollable passion: games of chance. Cards, dice, and chess became for a while his main source of income, but throughout the years—many years—they would rather be the cause of repeated set-backs: "I have squandered it all: my reputation, my time, and my money,"[10] Gerolamo bitterly regretted in the twilight of his life.[11]

In Padua, Gerolamo obtained his doctorate in medicine in 1526, after failing to do so in two previous examination sessions. And, presumably, his graduation was not so much due to Gerolamo's merit as to the resentment many professors nursed toward the brilliant student from Lombardy: brilliant but with a vain, coarse, and aggressive disposition, which led him to express his opinions and criticism in a too straightforward manner and adopt an antagonistic and offensive attitude.

> I am not very polite and I call a spade a spade; I let anger get the best of me to the point that later I regret it and feel ashamed. Among my shortcomings, there is a particularly nasty one: I cannot help myself from telling things that may be unpleasant for the other person to hear—actually, I enjoy it. I keep doing it, consciously and willingly, even if I know how many enemies I made because of it.[12]

After his graduation, in September 1526 Cardano moved to the town of Sacco (the present-day Saccolongo), near Padua, where he stayed for almost six years, practicing medicine. It was the "most carefree"[13] period of his entire life—"I played and listened to music, went for long walks, enjoyed lavish meals, and hardly did any work; I had no annoyances or anything to worry about, I was respected and honoured"[14]—even if occasionally disturbed by some unpleasant event. From time to time, Cardano would go to Venice to give vent to his passion for gambling. On one of those occasions, after two days of intensely playing cards at a private house, he realized that the cards were marked. A fierce scuffle followed, during which Gerolamo—who often carried one or more knives—stabbed his rival in the face, "although without excessive violence."[15]

He goes on:

> I took all the money, mine as well as his, and the clothes and rings I had lost the day before but had won back the next day. However,

seeing that my host was wounded, I willingly left some money; suddenly, I came face-to-face with the servants, who were unarmed. I listened to their pleas for mercy and promised to spare their lives if they opened the door for me. In the midst of all that confusion, their master, fearing the consequences of having cheated and realizing that he had little to gain from resisting, ordered them to open the door and so I could get away.[16]

The story had an unexpected sequel:

That very same day, at two o'clock at night, as I tried to avoid being arrested for having assaulted a senator and was carrying the knives hidden in my coat, I slipped and fell into water. While I was falling, I did not panic; with my left hand I grabbed the side of a boat and was rescued by friends. When I climbed aboard—this may be hard to believe—the man with whom I had played cards, his face bandaged due to his wound, offered me dry clothes, which I put on, and we arrived in Padua together.[17]

During his years in Sacco there were also much happier times for Cardano, as when he met a local young lady, Lucia Bandareni, and immediately began to "burn with love for her."[18] He also discovered, with the deepest relief, that he was "cured from the sexual impotence that had afflicted him throughout his youth."[19] They tied the knot in 1531, and from their union three children were born: Giovanni Battista in 1534, Chiara in 1537, and Aldo six years later.

In his memoirs, Cardano writes that he had a premonition of meeting Lucia in a dream, in which "in a charming and beautiful garden, adorned with flowers and fruits of all sorts," he had seen "a young maid dressed in white"[20] in whose features he had recognized his future wife. But this is only one of the innumerable oneiric visions described by Gerolamo, who believed he could predict the future through dreams thanks to the "extraordinary powers"[21] he claimed to possess. Another "miraculous faculty"[22] he professed to have, for instance, was hearing a humming sound every time someone, somewhere was talking about him: in his right ear if kind words were spoken, and in his left ear if something malevolent was being said. He was also convinced of possessing an "illumination power," a "kind of halo" that surrounded and protected him "from his enemies and imminent danger," and which helped him "to achieve experience, authority, profit, and good results in his studies."[23] In his autobiography, Cardano—fond of, among many other things, magic, horoscopes, divinations, and occultism—also claimed to hear voices coming from nowhere and see demonic things; he carried amulets and precious stones with mystical powers to chase away harmful influences, and also relied on a protective spirit.

In light of all this, observes historian of science and Cardano's biographer Oystein Ore, "it is no wonder that some of his contemporaries believed that he was not in his right mind."[24] In fact, plenty has been written about a possible mental pathology of the Lombard scholar, with many diagnoses emitted based on his writings. However, Ore goes on, "it seems difficult to find any convincing argument for a verdict of insanity, except possibly during the final period of his life,"[25] marked—as we shall see—by a series of dramatic events that might have significantly affected his mind. Besides, most of the accounts about supernatural incidents are found in his autobiography, written very late in life by an aging Cardano, who was suffering from a senile condition aggravated by the misfortunes he endured and a profound solitude. The suspected pathology, then, would have been limited to the period during which the text was written, and might not have been present when the reported facts actually took place.

In conclusion, although Gerolamo Cardano was undoubtedly a singularly eccentric personage, we must not forget that in those times superstition and belief in demons and evil spirits was the rule rather than the exception, and that a clear distinction between science and magic did not yet exist.

In February 1532, shortly after the wedding, Cardano decided with his wife's agreement to leave Sacco and move to Milan, given that his medical practice in the small Venetian town did not provide him with an adequate income to meet his new family responsibilities. And so, as he had unsuccessfully done three years earlier, he requested to be admitted to the city's College of Physicians, but his application was once again turned down. The reason was still the same: the College's regulations did not allow the admission of illegitimate children, and according to the members of the admissions committee, Cardano had to be considered as such, for his parents got married only after his birth. In order to practice his profession, at the end of April Gerolamo moved to the nearby town of Gallarate, where he spent nineteen months characterized by severe financial difficulties for lack of enough patients. Faithful to his old habits, he sought in dice and cards an alternative to make ends meet; but things did not work out for him on that front either. At a certain point he was obliged to pawn some furniture and his wife's jewels to pay up gambling debts.

In 1534, at the edge of the precipice, Cardano had a providential and decisive stroke of luck by becoming friends with the Milanese diplomat Filippo Archinto, thanks to whom he was appointed to the chair—formerly held by his father Fazio—of astronomy, arithmetic, and geometry in the Piattine Schools (established in 1503 in Milan through a bequest from the nobleman Tommaso Piatti). Little by little, thanks to Archinto's connections, Gerolamo got acquainted with prominent and influential members of the city's society; among these, Francesco Sfondrati, a future cardinal, and Captain Giovanni Battista Speciano, who would introduce Cardano to Alfonso d'Avalos d'Aquino d'Aragona, Marquis del Vasto, who in 1538 would become governor of Milan.

In June 1535, Cardano once again requested admission to Milan's College of Physicians, and just as before, his application was turned down; this time, however, he was granted the right to practice medicine in the city, albeit with certain conditions. It did not take long for Gerolamo's medical skills to earn him a good reputation—much to the envy of his colleagues—after successfully treating some members of Milan's aristocratic families, such as the Borromeos, and obtaining their valuable protection as a result. Still full of resentment at being excluded from the College, in 1536 Gerolamo published the controversial treatise *De malo recentiorum medicorum usu medendi libellus* (Satire on the Misguided Use of Medicine by Present-Day Doctors),[26] in which he harshly denounced and condemned what were in his opinion the recurring mistakes of the physicians of his time. Although he had already authored several texts, this was his first printed publication. In the course of his life, Cardano would produce more than 200 works—some of them of considerable size—"on every conceivable topic."[27]

As was to be expected, his provocative treatise triggered a disgruntled reaction from Milan's medical profession, but its author's reputation got a boost that reached beyond the city's walls, and not only in a negative way. Shortly after the publication of the book Cardano was offered a chair in medicine at the University of Padua, but he declined it "because it did not provide me with an adequate income."[28] In 1537 he failed for the nth time to gain admission to the College of Physicians, but his time would come two years later. Thanks to the pressure put forth by some of the city's influential personalities, in the summer of 1539 the College's policy was modified: sons who became legitimate following the union of their parents could also be admitted. And so, the last obstacle being removed—in fact, the rule had been changed expressly for him—he was admitted as a member of the College on August 14 of the same year, and went on to become Milan's most famous and sought-after physician. In 1541, those very members of the College who had for so long denied him admission now appointed him, willingly or unwillingly, as their new rector.

Toward the end of the 1530s Cardano—constantly pursuing his interest in the most diverse disciplines—had begun to write an extensive text in practical mathematics, intended to illustrate applications of arithmetic, algebra, and geometry using an approach based on Luca Pacioli's *Summa*. Regarding the theory of equations, Girolamo had decided to consider only those of first and second degree and avoid the treatment of cubic equations in view of Pacioli's well-known claim as to the impossibility of finding a general solution formula for the latter.

However, while he was working on the book—probably between 1537 and 1538—he received in Milan the unexpected visit of a strange personage: a tall, thin man, with olive-colored skin and eyes deep in their sockets, rather taciturn and clumsy in his movements, but giving the impression of being a great mathematician.[29] The man put Cardano on the spot by challenging him to solve some problems leading to third- and fourth-degree equations. Gerolamo did not yet know that his mysterious visitor, that unrepentant Messer Zuanne de Tonini da Coi, was in the habit of pestering other mathematicians with problems whose solution he did not know. But it wasn't long before he discovered it by himself. Messer Zuanne went on to tell him that, notwithstanding Pacioli's claim, someone certainly possessed the formula for cubic equations, in particular for the case "cube and things equal to number" $[x^3 + bx = c]$, and he also told him all about the sensational Venetian challenge that had taken place a few years earlier.

The news galvanized Cardano into action. It is plausible that he might first have tried to find the formula for the solution of cubic equations by himself, but without success. With his mathematical treatise soon going to press, and determined to include in it the new and extraordinary algebraic result, Gerolamo came to the conclusion that he should contact the discoverer of the formula directly. And so, he called a friend about to leave for Venice, the bookseller Zuanantonio da Bassano, and asked him to carry an important message from him addressed to Niccolo Tartaglia.

Messer Zuanantonio met Tartaglia in Venice on January 2, 1539. According to the transcription that appeared seven years later in Niccolo's *Quesiti*, the dialogue began with the bookseller's message:

Messer Niccolo, I was sent by a good person, a doctor in Milan named Messer Hieronimo Cardano, an eminent mathematician who gives public lectures on Euclid in Milan and is about to publish a treatise in arithmetic, geometry, and algebra that will be a wonderful thing. He has learned that you were involved in a contest with Maestro Antonio Maria Fior; that you both agreed to propose thirty questions each and so you did. His Excellency also heard that all thirty problems proposed by Maestro Antonio Maria led to the form thing and cube equal to number; that you had found a general rule for that case and that, thanks to the power of your discovery, you answered all his thirty questions in two hours. Therefore, His Excellency requests that you please send him the rule you found and, if you wish, he will publish it in his book under your name, but if you prefer not to make it public, he will keep it secret.[30]

After reading these words, one could not say that Cardano's request was unreasonable. After all, four years had passed since Tartaglia's discovery of the solution formula for the case "thing and cube equal to number," and he had no plans to publish anything in the near future that could have included the delicious fruits of his algebraic research. In fact, up until then he had published only the small treatise on ballistics, *Nova scientia*, a remarkable work, to be sure, but with no relation whatsoever to cubic equations. Cardano, on the other hand, had almost finished a mathematical book whose significance would dramatically increase with the inclusion of the valuable formula. However, whatever credit rightly belonged to the inventor of the formula, even if properly acknowledged, it was very likely—Tartaglia might have reasoned—-that the major benefits in terms of honors and prestige would be reaped by the author of the book. Besides, Niccolo had already revealed to Tonini da Coi that he really had no intention in "keeping secret" his discoveries but rather to "publish them for everyone" in due time. This is essentially what he told the bookseller as he rejected Cardano's request: "Tell His Excellency to forgive me, but should I decide to publish my invention I wish it to appear in one of my own works and not in someone else's; and beg His Excellency to excuse me."[31]

The conversation proceeded with a new request, followed by a new refusal:

> ZUANANTONIO: If you do not wish to reveal your invention, His
> Excellency asks you to please let him have at least the thirty
> problems that he [Fior] proposed to you, together with your
> solutions, and similarly the thirty problems you proposed
> to him.
> NICCOLO: I could not grant him such request either, because upon
> seeing one of these cases and its solution, His Excellency would
> immediately understand the rule I have found, with which many
> other similar rules could be discovered.[32]

At this point, and quite unexpectedly, Messer Zuanantonio handed Tartaglia a list of problems leading to equations of third and forth

degree that Cardano had drawn up to challenge the Brescian mathematician (these appear in square brackets written in modern notation).[33]

His Excellency has given me seven questions for you, and asked that
you please answer them. Here they are:

1. Divide 10 in four proportionally continuous parts
    (i.e., forming a geometric sequence, in modern terminology)
    such that the first one be 2.* $[x^3 + x^2 + x = 4]$

2. Divide 10 in four proportionally continuous parts such that
    the second part be 2. $[x^3 + x^2 + 1 = 4x]$

3. Find four proportionally continuous numbers such that the
    first one be 2 and the second and fourth added together make
    10. $[x^3 + x = 5]$

4. Find four proportionally continuous numbers such that the
    first one be 2 and the third and fourth added together make
    10. $[x^3 + x^2 = 5]$

5. Find four proportionally continuous quantities such that the
    second one be 2 and the first and fourth added together make
    10. $[x^3 + 1 = 5x]$

6. Make from 10 three proportionally continuous parts such that
    the first multiplied by the second makes 8.†
    $[x^4 + 8x^2 + 64 = 80x]$

7. Find a number such that multiplied by its [square] root plus 3
    makes 21.‡ $[x^3 + 3x^2 = 21]$.

* Let a, b, c, and d be the four numbers to be found. They are proportionally continuous if
$b : a = c : b = d : c$. The equation arising from the problem is obtained by calling x the common
ratio and imposing the given conditions ($a + b + c + d = 10$, and $a = 2$). Once the value of x is
found by solving the equation, setting $a = 2$, the values of b, c, and d are immediately obtained.
A similar approach may be used to solve problems 2 to 5.

† Let a, b, and c be the three numbers to be found. From the given conditions ($a + b + c = 10$;
$b : a = c : b$, and $a \times b = 8$) and by suitable algebraic operations, one obtains the equation
$b^4 + 8b^2 + 64 = 80b$. To be consistent with the notation used in the other problems, we have
denoted the unknown by b instead of x.

‡ Let x^2 be the number to be found and x its square root. The given conditions lead us to the
expression $x^2(x + 3) = 21$, from which the final equation follows by algebraic manipulations.

Can we reasonably assume that Cardano knew how to solve the equations resulting from the above problems, or at least most of them? No, certainly not. First of all, the equation arising from the third question is of the type "cube and things equal to number," and Gerolamo could not have known the solution formula: otherwise, why send someone to Venice to obtain it from Tartaglia? And that equation is not even the most difficult on the list. In view of all this, it was easy for Niccolo to see the truth behind the Milanese mathematician's subterfuge and to trace back the real source of the problems being proposed to him:

> These are Messer Zuanne da Coi's questions and no one else's, because I recognize the last two: he sent me one similar to the sixth two years ago,[34] and I made him admit that he didn't understand nor know how to answer it; and to one resembling the last one (leading to cenno and cube equal to number) I gave him by courtesy the solution less than a year ago,[35] from which he found a particular rule to solve similar problems."[36]

Hence, Cardano had merely passed on to Tartaglia the problems he had received from Tonini da Coi, a move hardly conducive to eliciting Niccolo's sympathy, despite Messer Zuanantonio's explanations:

> ZUANANTONIO: I know for certain that these questions were given to me by His Excellency Messer Hieronimo Cardano and no one else.
>
> NICCOLO: Therefore Messer Zuanne da Coi must have gone to Milan and posed the questions to His Excellency who, unable to answer them, sent them to me, and he now expects me to solve them. This is certainly the case, because last year Zuanne promised me to come to Venice but never did. I now believe he changed his mind and went to Milan instead.
>
> ZUANANTONIO: Do not think that His Excellency would send you these questions if he didn't understand or couldn't solve them, or if they belonged to someone else, because His Excellency is among the most learned in doctrine in Milan and the Marquis del Vasto[37] has largely rewarded him for his expertise.

NICCOLO: I do not deny that His Excellency is a most
knowledgeable and competent person. All I am saying is that he
does not know how to answer these seven questions he sent me
to solve with a general rule. Because if His Excellency is unable
to solve the one involving thing and cube equal to number
(which you so insistently tried to obtain from me) how could he
be able to solve most of the questions that lead to much more
difficult equations? And if he knew how to answer these, much
more easily would he know how to solve the thing and cube
equal to number case, and would not be begging nor looking for
the solution.[38]

At a loss for an argument to refute Tartaglia's, Messer Zuanantonio
asked him simply for the statements—without the solutions—of the
problems with which Niccolo and Fior had challenged each other. His
request was only partially granted:

ZUANANTONIO: I don't know how to reply because I'm not versed
in these matters, but if you could talk to him [Cardano], I think
he would know what to tell you. But let's put all these questions
aside; so that my visit would not have been in vain, let me at least
have a copy of the thirty cases Maestro Antonio Maria Fior
proposed to you. And if you could also give me a copy of the
thirty you proposed to him, you would make me extremely
happy.
NICCOLO: I shall give you a copy of his questions (even if I am
pressed for time), but I cannot give you mine because I haven't
got any copies, and I don't remember them since they were all so
different. But if you go to the notary, he will be able to give you
a copy.[39]

The visit ended after Tartaglia communicated Fior's questions to
Zuanantonio. The bookseller returned to Milan and reported the result
of his mission to Cardano. Not surprisingly, "His Excellency's" reaction
was far from enthusiastic. Without delay, on February 12, 1539, he wrote
Niccolo a letter full of anger, resentment, and stinging sarcasm. It was

the first personal contact between Cardano and Tartaglia. Decidedly, the relationship between two of the most important and emblematic scientific figures of their time did not get off on the right foot. The letter began as follows:

> I am greatly amazed, dear Messer Niccolo, at your discourteous response to a certain Zuanantonio da Bassano, bookseller, who was sent by me to ask you for the answers to seven questions and for a copy of those you and Maestro Antonio Maria Fior exchanged, together with their solutions. You did not deign to send me any of them, except those thirty of Maestro Antonio Maria, which are really all of the same sort: cube and thing equal to number. It saddens me, among other misfortunes afflicting this art [that is, algebra], that those who practice it are so disrespectful and presumptuous that, not without reason, people consider them almost crazy.
>
> Let me wake you up from this fantasy as I recently did Messer Zuanne da Coi, who believed he was the wisest man in the world and left Milan in disgrace. I would kindly like to make you realize that you are closer to the valley than to the top of the mountain.[40]

After this caustic start, Cardano's tone became much less hostile and even condescending toward the willful but surely competent recipient of the letter:

> It might be that in other domains you are more proficient and skillful than your response suggests. I assure you of my esteem and as soon as your book on ballistics[41] came out, I bought the only two copies brought by Zuanantonio, of which I gave one to Signor Marquis[42] and kept the other for me. I highly complimented you to Signor Marquis, believing you to be more grateful, kinder and polite than Messer Zuanne, to whom you consider yourself superior. But I see little difference between the two of you, unless you show me otherwise.[43]

Cardano goes on to say that he wishes to discuss and clarify four points, the first one regarding Tartaglia's accusation of having proposed seven questions that were in fact those of Tonini da Coi. Niccolo's allegation was certainly justified, and Gerolamo was having a hard time explaining himself:

> The first point is your claim that my questions are not mine but Messer Zuanne da Coi's, which amounts to saying that there is no one here in Milan capable of thinking them up. My dear Messer, men of talent are recognized not by the questions they ask, as you believe, but by the answers they give; you therefore sin by serious arrogance. Plenty of people in Milan could pose those questions, and I did before Messer Zuanne knew how to count to 10, if he is as young as he appears to be.[44]

The second and third points also regarded the questions Cardano had sent to Tartaglia. Gerolamo is again scornful and sarcastic, but his accusations—as we shall soon see—were probably due to a misunderstanding:

> You have told the bookseller that solving one of Maestro Antonio Maria's questions would solve all of mine. Let me please ask you: To whom do you think you are talking, to one of your pupils? Where did you learn that the rule for solving the equation "cube and things equal to number" [that is, $x^3 + bx = c$] could also be the solution of a question cube and number equal to thing [$x^3 + c = bx$]?
>
> I say the same thing about the other questions. In attempting to impress a bookseller with the mastery of your art, you made yourself appear a total ignoramus in the eyes of the experts. However, I do not consider you ignorant but rather too pretentious, just as Messer Zuanne da Coi who, seeking to make others believe he knew something he didn't know, resulted in others believing he didn't know the things he did.
>
> The third accusation is your telling the bookseller that solving one of my questions would be to solve them all, which is absolutely false. It is a veiled insult to say that what I believed were seven questions was really only one, for this would suggest a serious mental mistake on my part. If I were a specialist in algebra, I would wager 100 ducats that the seven questions couldn't be reduced to one, or to two or three questions.

If you accept the bet I will go to Venice expressly for that, and will make a bank deposit as warranty here in Milan if you accept to come here, or you could do the same in Venice and I'll go over there.[45]

Cardano then raises a severe but justified objection to a passage of Tartaglia's *Nova scientia*, as if he wanted to take revenge on Niccolo for his insinuations about the questions he forwarded him. Whatever the case, Gerolamo also acknowledged his appreciation for the Brescian mathematician's work: "The fourth remark is a too obvious mistake in your book on the new science of ballistics. Forgive me for wishing to correct you, since on the topic of ballistics, which is hardly your domain of expertise, you have succeeded in saying many beautiful things."[46]

At the end of the letter, Cardano proposes two new problems to Tartaglia (the statements will appear later), with conditions attached: "I am sending you two problems and, separately, their solutions; if you are unable to solve the problems, the messenger will give you the solutions. However, you will get them one at a time, so that you don't think I sent them in order to learn them and not to give them to you. But give him first your solutions, lest you claimed to have solved the problem when in fact you haven't."[47]

Finally, Cardano once again asks Tartaglia for the statements of the problems with which he had challenged Fior, and he closes the letter in a benevolent tone:

> Would you care to send me the questions you posed to Maestro Antonio Maria Fior? If you don't wish to send the solutions, keep them to yourself, since you are so secretive. And, after you receive the solutions to my questions—in case you were unable to answer them, and once you are convinced that my previous seven questions are all different—if it pleases you to send me the solutions to some of them, as a testimony of our friendship and so that I can appreciate your great talent more than for any other reason, you would make me extremely happy.[48]

Tartaglia's reply, dated February 18, 1539, was fairly long and detailed, and by no means compliant. After having stigmatized the "arrogant and insulting language" used by the "Most Excellent Messer Hieronimo,"[49] he replied point by point to the allegations against him. In particular, regarding the seven questions he received through the bookseller Zuanantonio, he didn't back up one inch:

> You say that I have told the bookseller that those seven questions you sent me were not yours but Messer Zuanne da Coi's, almost implying that no one in Milan could answer them. Let me tell you that what I said is true, that those questions were his because a year and a half ago he proposed to me one similar to the sixth (but using different words); and I made him admit here in Venice that he didn't understand it nor know how to solve it; and for these and other reasons I concluded that the questions were really his and he was sending them to me under your name; but when the bookseller assured me that they came from Your Excellency, I reckoned that Messer Zuanne had gone to Milan and posed the questions to you (as I still firmly believe) and that Your Excellency, not being able to answer them, passed them on to me.[50]

Having reiterated his (legitimate) convictions, Tartaglia went on:

> Secondly, you accuse me of having told the bookseller that solving one of Maestro Antonio Maria's questions would solve all the seven you sent me.
>
> Thirdly, you also accuse me of having told the bookseller that solving one of your seven questions would solve them all; you added that this is absolutely false, and that you were willing to wager 100 ducats to prove your point. Regarding your second and third accusations, I reply to you that I believe you have dreamt those tall stories.[51]

Indeed, going over Tartaglia's conversation with Messer Zuanatonio it is clear that the points raised by Cardano were not convincing, and probably due to a misunderstanding by the bookseller or Gerolamo himself. In any case, Niccolo expressed again his disagreement, without backing away from Cardano's bet:

> It's absolutely true that I told the bookseller that Your Excellency would not know how to solve those seven questions with a general

rule, and to support my reasoning I told him that if you were unable to solve the case thing and cube equal to number $[x^3 + bx = c]$—which you so eagerly tried to obtain from me—even less could you solve your seven questions, which lead to more complex equations. And that if you were able to solve those more difficult cases, much more easily would you solve the case thing and cube equal to number. This is what I told the bookseller. But from what I gather, Your Excellency fervently wishes to show off his superiority; even if I were certain to lose, I would not refuse your invitation to deposit 100 ducats, and I will travel to Milan if you do not wish to come to Venice.[52]

Tartaglia then tried to reply to Cardano's peremptory criticism of his *Nova scientia* by first asserting—with dubious subtlety—that "your arguments are so weak and badly stated that a sickly woman could knock them down."[53] Niccolo's explanations, however, were somewhat shaky and confusing—Cardano's criticism, let us repeat it, was well founded—and he concluded his attempted defense in these terms:

> With your ridiculous antagonism you sought to demonstrate how exceptional you are, but in fact you only proved to be, I wouldn't say a complete ignoramus, as you called me, but a man of poor judgment.
>
> Your Excellency says that he excuses my mistakes, since ballistics is not my domain of expertise and yet I succeeded in saying some excellent things about that art.
>
> In this respect, I say that I find delight in new discoveries, and in studying and talking about things that no one has studied or discussed; and I don't take any pleasure, as so many do, in filling up pages with things stolen from this or that author. And even if talking about artillery fire is not something very honourable in itself, I found it worthy of study because the subject is new and interesting.[54]

The next passage of Tartaglia's letter is very significant. It is clear that despite his distrust Niccolo must begin to realize the advantages of a more friendly relationship with Cardano, who enjoyed the esteem, the friendship, and the protection of the Marquis del Vasto and governor of Milan Alfonso d'Avalos. Tartaglia could have learned this from the bookseller Zuanantonio and later from Gerolamo's letter. Besides being "one of the most powerful men in Italy,"[55] Governor d'Avalos was also known for his intellectual sensibility and for the uncommon and extreme generosity toward the scholars with whom he enjoyed associating. Therefore, "through Cardano, Tartaglia could be introduced to the Milanese court," and would have a chance "to explain in person to the governor his inventions in artillery and possibly to obtain a well-paid position as a technical expert."[56] It was probably with such thoughts in mind that Niccolo—after having replied tit for tat to Cardano's accusations—changed the tone of his letter and adopted a more receptive and helpful attitude in announcing gifts for the governor and Cardano—two geometrical instruments of his own invention.

> I have fabricated two kinds of instrument to be used in this art of ballistics: a triangle to adjust artillery fire and evaluate heights, and another one to evaluate horizontal distances.
>
> And because you told me that you bought two of my books, one of which you gave to Signor Marquis and the other you kept for yourself, I shall send you four such instruments. I have given them to Signor Ottavian Scotto,[57] who will arrange their delivery to you. Of the four instruments, two are for His Excellency Signor Marquis and the other two for you.[58]

Tartaglia then proceeded to discuss the two new questions Cardano had included in his letter:

> The first question: Divide 10 in four proportional continuous parts such that their squares added together make 60. Fra Luca [Pacioli] asks a similar one but does not answer it.
>
> The second one: Two men possessed together I don't know how many ducats. They earned the cube of one-tenth of their capital; had they earned 3 less than they did, they would have earned as much as the amount of their capital. Find their capital and the sum they earned.[59]

While he elegantly solved the first problem by resorting to some simple quadratic equations, Niccolo did not attempt to solve the second one, which leads to the cubic equation $1000x + 3000 = x^3$, of whose general form $bx + c = x^3$ ("thing and number equal to cube") Cardano certainly did not possess the solution formula.* Therefore, neither could he know the solution of this particular problem.[60] To sum up: in his eagerness to extract the secret rule from Tartaglia, or at least glean a few clues, he had again resorted to trickery, and not very cleverly. Tartaglia's rebuke was predictable, but this time in a tone that might be considered good-natured:

> Regarding your second question, I had a good laugh, for I see that Your Excellency wishes to play *Trappola*[61] [Trap] or *Corrigiola*[62] with me, as the gypsies do, and he thinks I can be fooled by the promise to send me the solution if I don't know how to solve it.
>
> Using algebra, this problem (as you are certainly aware) leads to the case things and number equal to cube, and I maintain that you don't know the rule for solving it. And I'm so certain of what I say that I'm willing to bet ten ducats against one.[63]

After offering the bet, Tartaglia explained the reasons for his incredulity about Cardano's ability to solve his own problem:

> I say that after having found the rule for the case thing and cube equal to number $[x^3 + bx = c]$, thanks to some insights from that discovery the next day I also found the general rule for things and number equal to cube $[bx + c = x^3]$, which I could never have discovered without the first one. And because you don't know such rule, even less would you know the one for things and number equal to cube, which you thought you would obtain from me by cunningly pretending you had given the solution to the messenger. Such lie made me doubt whether you even know the answer to your first question, and which I am sending in this letter.[64]

---

* Denoting by x the capital, the profit is then $(x/10)^3$. The given conditions lead to the expression $(x/10)^3 - 3 = x$, from which the final equation follows after some algebraic manipulations.

Then, partially fulfilling one of Cardano's requests, Tartaglia revealed the statements—but not the solutions—of some of his questions to Fior during their Venetian duel; besides those he had disclosed to Tonini da Coi (listed in the previous chapter) there were some in plane and solid geometry. He also reported on his activities in Venice, and showed again his deference toward his fellow mathematician:

> You asked me to please send you the questions I posed to Maestro Antonio Maria Fior, and should I not wish to send their solutions, to keep them for myself. It would be rather long for me to write down all thirty questions. Presently I am busy putting up some public posters, because next Sunday I shall begin lecturing at San Zuanepolo[65] on the science of weights and on the practical application of certain things I discovered about artillery fire and various other topics. Not wishing Your Excellency to believe this to be a pretext for not sending my thirty questions, I am enclosing a copy of the poster I put up two days ago. And to show you my willingness to serve you (even if I am busy) I am sending you in the meantime nine questions that I still recall; for I honestly did not write my questions down and don't have them at hand. But as soon as my public lectures are over I shall fetch a copy from the notary and send it to you.[66]

Tartaglia ends his long letter by refusing to send the solution to any of the seven questions he received through the bookseller Zuanantonio, convinced that Cardano was unable to answer them. But in fact, there was another good reason for his refusal: Niccolo didn't know how to solve many of them himself!

Cardano replied to Tartaglia on March 19, 1539. The tone of his letter is clearly set from the start:

> Messer Niccolo, my dearest. I have received your very long letter; the longer it went on, the more I enjoyed it, and I wished it had been twice as long. Do not take my harsh words in the wrong sense; they are not inspired by hatred—there is no reason for that—or by a malevolent disposition: whenever I can I act righteously, as I am obliged to do in the exercise of my medical profession. Neither am

I driven by envy, because if you were equal or inferior to me, it would be pointless, and if you are superior to me in this art [algebra] I should be seeking to emulate rather than discredit you. Besides, he who is envious speaks ill in the absence, not in the presence of his target.[67]

From beginning to end, Cardano's letter is full of warmth and flattery. In particular, after such a friendly start Gerolamo wants to explain the reasons for the "harsh words" he used in his previous letter, clarifying the nature of his relationship with Tonini da Coi and hence conceding—if not altogether explicitly—what Niccolo had surmised about the origin of the seven questions:

I wrote those things to provoke you and prompt you to reply, for I deemed you to be a person of exceptional intelligence, judging from Messer Zuanne da Coi's account when he came here. I assisted and pleased him as best I could, and I even contemplated giving him one of my works. But he behaved ungratefully, saying wicked things in public and in private, and deliberately challenging me with posters and writings. Things not having turned out to his liking, [. . .] he left in desperation abandoning a school of some sixty students, which saddened me greatly.

If I have written harshly to you I did it on purpose, expecting to provoke what in fact happened: a reply from you with the friendship of a person unique in this art [algebra], as I could infer from the things you wrote in your letter; and so I don't regret having behaved in such inappropriate way.[68]

Girolamo then acknowledges reception of the poster announcing Tartaglia's lectures on statics and ballistics in Venice and says he is still waiting for the instruments Niccolo had promised to send him and the Marquis d'Avalos. He then mildly mentions a couple of questions that had been a source of friction between them, trying to erase any trace of controversy and lavishly praising the Brescian mathematician. He especially expresses his desire to welcome him in Milan, adding that the marquis had urged him to write to him without delay:

Regarding the response to my four accusations, I shall only reply to two of them. One is about your new art [i.e., *Nova scientia*]; the other concerns your claim to being the most knowledgeable in this art [algebra]. With respect to the second one, I'd rather live a lazy life than die a gifted man,[69] especially since you have already dismissed it by saying that Zuanantonio had misunderstood you; so I declare this dispute settled. I hope you will come to Milan to meet me without the deposit of one hundred ducats, because I consider you an intelligent man, and after becoming acquainted with each other we will be able to come to a decision.

As for our disagreement [over *Nova scientia*], you are certainly right in defending your work already published. And of course, when you come to Milan (as I hope, God willing) we will be more at ease to discuss it.

Yesterday evening I received your letter, and today I am writing to you as requested by Signor Marquis; for this reason I was not able to go over your other propositions. In any case, please send me or bring with you the rest of the questions you proposed to Maestro Antonio Maria."[70] Cardano then once again asked Tartaglia for the one thing he cared most about at that moment: the solution formula for cubic equations. He did it by repeating the reasons Niccolo had already heard from the bookseller Zuanantonio:

> If you would send me or bring with you some of the solutions obtained by using your rule, it will give me an immense pleasure, because I appreciate a kind gesture, and I have written a text on the practice of geometry, arithmetic, and algebra of which more than half have so far been printed. If you'd care to give me your rules, I'll include it under your name at the end of the book, as I have done with all the others who gave me something interesting, and I shall mention you as the inventor; but if you prefer that I keep them secret, I'll do as you please.[71]

It would be natural to suspect that Cardano's friendly attitude toward Tartaglia in this letter was, after all, nothing other than a deliberate strategy on his part to be in Niccolo's good graces and get hold of his formulas. This could be true, at least in part. But we must not forget that Cardano's was a response to Tartaglia's own letter in which, despite some expressions of criticism and disapproval, he was clearly attempting to establish a more balanced and cordial relationship with Gerolamo—a gesture that must have been greatly appreciated by him. In conclusion, it might not be far from the truth to say that, after the initial misconceptions and the persistence of some degree of mistrust (especially on Niccolo's part), a bond of sincere esteem and admiration developed between the two mathematicians. Each respected the other as a scholar of remarkable caliber and realized the mutual advantages that could stem from their friendship.

In closing his letter dated March 19, 1539, Gerolamo again invited Tartaglia to travel to Milan for the purpose of introducing him to the Marquis d'Avalos:

> I informed His Excellency Signor Marquis about the instruments you sent him (which have not yet arrived) and about the poster, which His Excellency asked me to read for him; all these things captivated His Excellency. That is why he urged me to write you this letter on his behalf, advising you that you must absolutely come to Milan, and that he would like to talk with you. I thus exhort you to come straight away, without hesitation, because Signor Marquis is so generous in rewarding men of learning and talent, so liberal and magnanimous that no one who served His Excellency has ever been displeased. So do not fail to come, and stay over at my place. That is all. May Christ protect you from evil.[72]

The letter produced the desired effect, and a few days later Tartaglia arrived in Milan as Cardano's guest. The Marquis was momentarily out of town, in Vigevano, some thirty kilometers away. While they awaited his return, Niccolo and Gerolamo had plenty of time to discuss and debate on various topics, in particular the question of cubic equations and their solution formulas.

In his 1546 *Quesiti*, Tartaglia transcribed the highlights of his discussion with Cardano on this subject, which took place on March 25, 1539:

> GIROLAMO: I am very happy you have come while His Excellency Signor Marquis is in Vigevano, because we will be at ease to talk

about things that interest us until his return. It was discourteous on your part not sharing with me the rule you found for the equation thing and cube equal to number [$x^3 + bx = c$], especially since I had so insistently asked you for it.

NICCOLO: It is not about the rule or the things that can be found using it that I am secretive, but because of the discoveries that can be made from it; it is a key that opens the path to finding the solution to infinitely other equations. And if I weren't busy with the translation of Euclid into the vernacular, I would have already found the general solution to many other equations.

But as soon as I've finished translating Euclid, I'll write a practical textbook and, together with it, a new Algebra, in which I intend to publish all the new rules I've found, as well as others I hope to discover. And I also wish to show the rule from which infinitely others can be found, which I expect to be a useful and interesting thing.

This is why I refuse to reveal my formulas, even if at present I care little about them (being busy with Euclid, as I told you). If I showed them to some scholar such as Your Excellency, he could easily find other solution rules (for it's easy to add to what is already known) and publish them claiming to be their discoverer. In so doing he would spoil all my projects. This is why I have been so discourteous with Your Excellency, who is presently having his work on a similar topic printed, and has written to me about publishing my formulas under my name and mentioning me as the inventor. That is something that really displeases me, because I prefer to publish my discoveries in my own works, and not in someone else's.

GEROLAMO: But I have even written to you that if you didn't wish to disclose them, I would keep them secret.

NICCOLO: I can only say that in that respect I didn't want to believe you.[73]

Then, to convince Tartaglia of his commitment to keep secret any discoveries that might be revealed to him, Cardano took the following oath:

I swear, on the Sacred Gospel of God and as a true gentleman, not only never to publish your discoveries, if you ever showed them to me, but I also promise, and engage my Christian faith, to write them down in cipher, so that after my death no one could understand them.[74]

Having taken his oath, and given that there was nothing more he could say, Gerolamo added: "Believe me if you wish; if you don't, forget about it."[75] And Niccolo, after having resisted for so long, began to yield:

NICCOLO: If I don't believe your pledge, I would rightly deserve to be called a man without faith. I've decided to ride to Vigevano to meet His Excellency Signor Marquis, because I've been here three days already waiting for him. On my return, I promise to show you everything.

GEROLAMO: Since you are going to meet Signor Marquis, I'll give you a letter of introduction for His Excellency so that he knows who you are. But before you leave, I would like you to show me the solution formula for those equations, as you promised me.[76]

At that point, Tartaglia gave in. After having strenuously defended it for many years, he decides to reveal his secret formula by showing Cardano his three solution rules for the third-degree equations "cube and thing equal to number" $[x^3 + bx = c]$, "cube equal to thing and number" $[x^3 = bx + c]$, and "cube and number equal to thing" $[x^3 + c = bx]$. However, Tartaglia does not write down the rules using symbols—his algebra, remember, did not possess any yet—but in words; more precisely, in the form of a poem. A kind of riddle in rhyme, rather cryptic, the reason for which Niccolo explained to Cardano:

You should know that, in order to remember the sequence of operations each time I need them, I have written them down in rhyme, because without such precaution I would have often forgotten them.

And I don't care if my words in rhyme are not easy to understand, because they are enough for my purpose, which is to recall the rule whenever I need it. I shall write it down for you in my own hand, so that you are certain it is the correct one.[77]

Here is Niccolo Tartaglia's algebraic poem:

When the cube with the things together
Are set equal to some discrete number [i.e., an integer]
Find other two that by it differ.
You will get into the habit
Of making their product always equal
To the third of the things cubed exactly,
The result then
Of their cubic roots properly subtracted
Will be equal to your principal thing.
In the second of these cases
When the cube remains alone
You will take these other steps
From the number you will make two parts
So that one in the other yields
The third of the cube of the number of things
Of these two parts by known precept
You'll joint together their cubic roots
And this sum will be your answer.
Then the third one of your calculations
It's solved with the second, if you look carefully
Because by nature they're related.
All these I found, and not with slow steps
In fifteen hundred thirty four
With firm and vigorous foundations
In the city surrounded by the sea.[78]

Once the formula was revealed, the meeting came to an end. The two mathematicians exchanged greetings, one of them requesting, and the other assuring him, absolute discretion:

NICCOLO: The solution rule is so clear that without any other examples I believe Your Excellency will completely understand it.

GEROLAMO: I almost did already. Go back to Venice, and I shall let you know if I have really understood it.

NICCOLO: And remember, Your Excellency, the promise you've made. Because if by any chance you would go back on it and publish my formula, whether in your forthcoming book or in some other one, and even if you do it under my name and mention me as its inventor, I promise and swear to you that I shall immediately put out another one that you will not appreciate at all.

GEROLAMO: I shall honour my promise, you can be sure of it. And please give Signor Marquis this letter from me.

NICCOLO: And now, I shall be on my way.

GEROLAMO: Go, and have a safe journey.[79]

Tartaglia then took leave, not only of Cardano, but also of a very young student and assistant of his, a seventeen-year-old from Bologna called Ludovico Ferrari, who had been present and listened to the conversations between the two scholars with keen interest.

Niccolo mounted his horse and rode out of Milan, but not in the direction of Vigevano. He went back to Venice instead, deliberately missing an opportunity to meet the illustrious Marquis d'Avalos, perhaps regretting already that moment of weakness in which he revealed the formula he had so long kept secret.

Cardano, for his part, began to study the formula right away. He had obtained it at last; now he had to understand it.

CHAPTER 5

The Old Professor's Notebook

On April 9, 1539, two weeks after their Milan meeting, Gerolamo Cardano wrote to Niccolo Tartaglia again. The short letter, whose tone was extremely friendly, began with Gerolamo expressing his surprise at Niccolo's sudden return to Venice after leaving Milan, and regretting his missed visit to the Marquis Alfonso d'Avalos:

> My dearest Messer Niccolo, your leaving so suddenly without having talked to Signor Marquis astonished me. He came on Holy Saturday[1] but could not collect your gift until Tuesday after Easter, and with great difficulty. He very much appreciated your instruments and would like to learn how to use them, so I briefly explained it to him. I certainly think it was a mistake on your part not to meet His Excellency, because he is a great admirer and supporter of talent and virtue. That's all for the moment. There might still be another occasion for you to make acquaintance with Signor Marquis.[2]

After mentioning the forthcoming publication of his mathematical treatise, Cardano resigned himself to admitting that he had not fully understood the poetic formula for the solution of cubic equations (this was due, as we shall see, to an inaccuracy in the wording of the formula),[3] and he asked Tartaglia for an explicit solution of the equation $x^3 + 3x = 10$:

> Regarding your solution method for thing and cube equal to number, I am grateful for having shared it with me, and you may be assured that I shall be indebted to you for that. But I must admit I have failed to understand it; and so I beg you, for the affection you feel for me and the friendship that binds us and which I hope will last as long as we live, to send me the solution to the following equation: a cube plus three things equals 10. I hope that you would be as happy to send it as I will be to receive it.[4]

The letter was signed with a warm "Yours very truly Hieronimo Cardano, physician."[5]

Tartaglia's reply, dated 23 April 1539, was much less cordial, but nevertheless offered to provide Cardano the clarifications he requested. The letter was addressed to "Honourable Messer Hieronimo"[6] and signed simply with the given name and surname of its author. Niccolo began by explaining the reasons for his abrupt return to Venice after the Milan meeting: "The cause of my so suddenly and silently leaving Milan without talking to His Excellency Signor Marquis is this: when I left Venice to go to Milan I promised my friends to be back for Easter, and if I had delayed my departure I would have been late, for I barely managed to get here by Holy Saturday."[7]

One may wonder whether the promise he made to his friends to spend Easter with them was really the only reason behind Tartaglia's decision not to go to Vigevano, and thus miss a valuable opportunity to meet the governor of Milan. In a passage of his *Quesiti*, Niccolo confesses having left Cardano's place distressed and devoid of enthusiasm, hardly in the mood for such a demanding visit and, instead, anxious to get home: "Upon my word, I do not wish to go to Vigevano; I'd rather go back to Venice, come what may."[8]

The cause of such state of mind appears obvious: Tartaglia had immediately regretted having revealed his secret formula to Cardano. In his April 1539 letter he declares himself anxious to read Gerolamo's mathematical treatise, to make sure he had not included the fateful formula and thus reneged on his oath. "As for your mathematical book,

I strongly wish it would come out soon so I could see it; because I fear you might go back on your word and reveal my formulas."[9]

Niccolo then wrote down the solution of the equation $x^3 + 3x = 10$, which Cardano had so eagerly requested, together with a detailed explanation, after having identified the line of the poem that had put Gerolamo off track.[10] Written in modern notation, the solution is:

$$x = \sqrt[3]{\sqrt{26}+5} - \sqrt[3]{\sqrt{26}-5}$$

To illustrate the use of the rule Niccolo also included the solution of the equation $x^3 + x = 11$,[11] and he closed his letter with a terse "Remember your promise."[12]

At this point, Cardano saw the light. As he fully grasped the meaning of Tartaglia's tercets, he became aware of their extraordinary power. He was in awe before the ingenuity of him who had succeeded in reaching a goal until then considered incredibly difficult, if not impossible.

And now, let us share Cardano's wonderment by paraphrasing Tartaglia's mathematical poem in modern algebraic symbols. To begin with, we shall examine the first nine verses, which describe how to solve cubic equations of the form $x^3 + bx = c$, that is, "cube and things equal to number":

When the cube with the things together	$x^3 + bx$
Are set equal to some discrete number [i.e., an integer]	$= c$
Find other two that by it differ.	$u - v = c$
You will get into the habit	
Of making their product always equal	$u \times v =$
*To the third of the things cubed exactly,**	$(b/3)^3$

* This is the line of the poem that misled Cardano: to specify the operation $(b/3)^3$, Tartaglia should have written "cube of one-third of the things exactly," rather than "the third of the things cubed exactly," which indicates the operation $b^3/3$. Here are the first six lines in the original Italian—it is the sixth that caused Cardano's confusion:

Quando che'l cubo con le cose appresso
Se agguaglia a qualche numero discreto
Trovan dui altri differenti in esso.
Da poi terrai questo per consueto
Che'l lor produtto sempre sia eguale
Al terzo cubo delle cose neto,

Then the result
Of their cubic roots properly subtracted* $\sqrt[3]{u} - \sqrt[3]{v}$
Will be the value of your principal thing.[13] $= x$

Tartaglia therefore reduced the solution of the equation $x^3 + bx = c$ to finding two numbers, u and v, that satisfy the conditions:

$$u - v = c$$
$$u \times v = (b/3)^3$$

Solving this "system" of two equations may appear a rather difficult task to a nonspecialist, but for sixteenth-century mathematicians it was not a particularly arduous one (and the same should be true for any present-day university student). Carrying out some simple algebraic operations on the system leads to the second-degree equation $v^2 + cv = (b/3)^3$, called *quadratic resolvent*, from which the value of v is obtained, and then, easily, that of u.[14] As historian Bortolotti has rightly observed, the quadratic resolvent of cubic equations is "one of the fundamental principles of the theory of algebraic equations," and its discovery was a crucial contribution to algebra on Tartaglia's part, given that "it was the foundation for the work of all subsequent algebraists."[15]

After the required calculations, the values of u and v are:

$$u = \sqrt{(c/2)^2 + (b/3)^3} + c/2 \qquad v = \sqrt{(c/2)^2 + (b/3)^3} - c/2$$

According to Tartaglia's specifications $x = \sqrt[3]{u} - \sqrt[3]{v}$. Hence the final formula for the solution of the equation $x^3 + bx = c$ is:

$$x = \sqrt[3]{\sqrt{(c/2)^2 + (b/3)^3} + c/2} - \sqrt[3]{\sqrt{(c/2)^2 + (b/3)^3} - c/2}$$

This is the secret rule Cardano coveted so much: the "invention" that had allowed Tartaglia to win the Venetian challenge against Antonio Maria Fior, bringing him fame and prestige among the scholars of his time.

---

* The two numbers (the cubic roots of u and v) must be subtracted so that the result is a positive number.

Once in possession of the above formula, Tartaglia—as we already know—had quickly obtained similar rules for the solution of the other types of third-degree equations appearing in his poem. The next nine verses concern the cubic equation of the form $x^3 = bx + c$, that is, "cube equal to things and number":

In the second of these cases	
When the cube remains alone	$x^3 = bx + c$
You will take these other steps	
From the number you will make two parts	$u + v = c$
So that one in the other yields	$u \times v =$
*The third of the cube of the things exactly**	$(b/3)^3$
Of these two parts by known precept	
You'll joint together their cubic roots	$\sqrt[3]{u} + \sqrt[3]{v}$
And this sum will be your answer.	$= x$

By a process similar to the one used in the previous case we find that the solution of the equation $x^3 = bx + c$ is:

$$x = \sqrt[3]{\sqrt{(c/2)^2 - (b/3)^3} + c/2} - \sqrt[3]{\sqrt{(c/2)^2 - (b/3)^3} - c/2}$$

The next tercet of Tartaglia's poem:

Then the third of our calculations
It's solved with the second, if you look carefully
Because by nature they're related.

refers to the cubic equation $x^3 + c = bx$, that is, "cube and number equal to things," which can be solved by reducing it to the previous case.[†]
Finally, the closing verses of the poem

---

* "The number of things" indicates the coefficient b of the unknown x. This line also contains an error: rather than "the third of the cube of the things exactly," Tartaglia should have written "the cube of one-third of the things exactly."

[†] Later, Cardano himself would show that if $x^3 + c = bx$ and $y^3 = by + c$, then

$$x = \frac{y}{2} \pm \sqrt{b - 3(y/2)^2}$$

These I found, and not with slow steps
In fifteen hundred thirty four
With firm and vigorous foundations
In the city surrounded by the sea.

inform us that these rules had been found in rapid succession ("not with slow steps") in 1535 (1534, according to the calendar of the Republic of Venice),[16] in Venice ("the city surrounded by the sea"), and established in a solid and indisputable way ("with firm and vigorous foundations"). In fact, Tartaglia never disclosed the proof of his formula, which, however, may be concealed in the verses of his poem.[17]

On May 12, 1539, Cardano sent Tartaglia an unbound and freshly printed copy of his mathematical treatise, accompanied by a polite letter in which he reiterated his deep appreciation and gratitude, besides expressing his sincere friendship and assuring Niccolo he had honoured the oath made in Milan:

> Regarding your letter of 23 April, which I did not receive until the day before yesterday, my dearest Messer Niccolo, let me reply to it point by point, beginning with your apologies for not going to Vigevano. I only wish what you wish, and I regret that you had to go through all that trouble because of my friendship without reaping any profit.
>
> To dispel any suspicion I'm sending you a copy of my book; it is not bound because it was freshly printed. I'm particularly grateful for your formula and your solution to my problem, which I appreciate more than if you had given me one hundred ducats. I admire your ingenuity, which surpasses that of everyone I've known. I consider you as the true friend you have proven to be.[18]

"As for your fears that I might print your invention," Cardano went on, "the oath I made to you should suffice."[19] Besides, Gerolamo argued, his having sent a copy of the book did not prove anything, since it would still be possible for him to add a new chapter in subsequent editions. Cardano absolved Tartaglia of his mistrust given the importance of the question (the "dignity of the matter"), but pointed out the error, because the only guarantee could come not from those printed sheets, but from "faith in a gentleman's word." His point was "that there is no greater betrayal than failing to keep your word and displease him who has pleased you; if you put me to the test, you would see whether or not I'm your friend, and whether I deserve your friendship and the pleasures that came with it."[20] Gerolamo ended the letter with a request: "I kindly ask you: regarding my books, out of affection toward him who has printed them and will sell them, to see that they get as wide a diffusion as possible. I would not ask you this if I had paid for the printing myself, because I care more about my friends' well-being than about my own."[21]

Cardano's voluminous text was titled *Practica arithmetice et mensurandi singularis* (The Practice of Arithmetic and Simple Mensuration).[22] It had been published "at their risk and expense"[23] by an editor and a printer in Milan, Bernardo Calusco and Antonio Castiglioni, who had paid the author the lump sum of ten ducats as remuneration for his work. The first page contained an image of Gerolamo, pictured as a man looking "young, glowing with health" and "with a determined face framed by a frizzy beard."[24] Later, in his autobiography, the scholar from Lombardy would give the following detailed description of himself:

I am of mediocre height. I have small feet, wider at the ends and curved; it's really difficult to find shoes my size and in the past I had them made to measure. My chest is rather narrow, my arms are slender and my right hand meatier, with stubby fingers—according to the necromancers, I should have been boorish and stupid: they should be ashamed of their alleged knowledge.

My line of life is short; the Saturnine line long and deep. My left hand is pretty, with well-shaped, tapering fingers; the nails are shiny. My neck is long and thin. I have a divided chin and a swollen and hanging lower lip. My eyes are very small and tend to half-close unless I watch intently. My forehead is large and hairless at the temples; my hair and beard were blond. I keep my hair well cut and my beard (which was, just as my chin, divided) short, thicker and longer under my chin.

I speak loudly, and those who pretended to be my friends reproached me for this habit; my voice is harsh and strong, but it didn't carry very far when I was teaching. I talk little and without much grace; my look is fixed, like that of someone who is reflecting.[25]

Divided into some seventy chapters, *Practica* presented various applications of arithmetic, geometry, algebra, astronomy, astrology, and land surveying. In particular, special attention was given to the fundamental arithmetic operations with integers, fractions, and irrational numbers; mnemonic rules for executing rapid calculations; questions in astrological numerology; and commercial, social, and agronomical topics (bookkeeping, determination of weights and measures, currency conversion, and calendar calculation). There was of course an extensive algebra section devoted to the theory of equations, but only to those of the first and second degree. Cubic equations were not mentioned, and therefore no reference was made to Tartaglia's formula. In keeping the secret, Gerolamo had also kept his word.

Reassured of the Milan mathematician's loyalty, Niccolo replied with a letter dated May 27, 1539, in which he began by thanking Cardano for sending the book, and adding that he had not yet examined it properly due to his poor health condition and his urgent work on the translation of Euclid's *Elements* into Italian (which he had already mentioned in a previous letter): "Most Honorable Messer Hieronimo, I have received your letter together with your book, for which I thank you. Not having been able to properly read it—since I'm very busy with Euclid's translation and not feeling well—I did nevertheless take a quick look at it."[26]

However, that "quick look" proved enough to enable Tartaglia to severely criticize some of Cadano's arithmetic calculations. In his letter, his disapproval takes the form of caustic comments ("Your Excellency's mistake is so flagrant that I'm left astonished, because anyone taking a glance would have spotted it"), and emphatic remarks ("absolutely false" and "extremely ridiculous"), regarding a certain rule for the extraction of cubic roots, later referred to as "something very far from the truth," so much so that—Niccolo concluded, somewhat compassionately—"I feel very sorry for the harm to your reputation."[27]

Leaving apart the harshness of such (wildly exaggerated) remarks, the result of a caustic disposition as well as the desire to demonstrate a supposed mathematical superiority, Tartaglia must have felt reassured, as he went over the pages of Cardano's work, not to find the slightest reference to the formula for cubic equations. It is not by chance that this time, when he signed the letter, he appended to his given name and surname an eloquent, friendly "All yours."[28]

Nonetheless, Tartaglia's touchiness forcefully surfaced a few months later, when he learned—in a July 10, 1539, letter from Mafio Poveiani, one of his students from Bergamo—about some rumors that had been circulating in Milan: "I have some news for you: one of my friends in Milan has informed me that doctor Cardano is writing another book about certain recently obtained results in algebra. I believe these are the things you told me to have showed him, and so I'm afraid he wants to trick you."[29]

Worried, if not terrified, Niccolo immediately replied to his pupil, asking him to keep an eye on the situation and let him know of any developments:

> Master Mafio, my dearest, I shall briefly reply to your letter. Regarding the news you have heard about doctor Cardano from Milan, it has certainly annoyed me; because if it's true that he wants to publish new results, these cannot be but those you mentioned. The proverb is right: what you don't wish to be known, don't tell anyone. Keep your eyes open: if you hear anything else on this matter, let me know.[30]

In fact, the rumors that Cardano was at the time writing an algebra treatise "on certain new results" would prove unfounded. It is possible, however, that Gerolamo had already achieved some significant progress in the field of cubic equations, helped by his own mathematical insight and his extremely precious collaboration with his Bolognian student Ludovico Ferrari, who had so attentively listened to the conversation between his teacher and Tartaglia in Milan.

Even if we cannot be certain, Cardano had perhaps already found a method for solving *all* third-degree equations, whatever their form, taking advantage of and going beyond the results already obtained by Tartaglia. However, he had yet to overcome some serious obstacles.

While he was studying one of the types considered by the Brescian mathematician, $x^3 = bx + c$ ("cube equal to things and number"), Cardano realized that the solution rule didn't seem to work, since it led to meaningless results for particular values of the coefficients b and c. In a letter dated August 4, 1539, Gerolamo informed Tartaglia of the problem, first regretting not having received an answer to his previous correspondence:

> You have not yet deigned to reply to several letters I sent you, and in which I asked you for the answer to various questions, that of cube equal to things and number among these. I'm sure to have understood the rule, but when the cube of one-third of the things $[(b/3)^3]$ is larger than the square of half the number $[(c/2)^2]$, I cannot apply it such as it is. Therefore, I would be grateful if you could solve the following equation for me: a cube is equal to nine things plus ten $[x^3 = 9x + 10]$. You would make me immensely happy by granting me this.[31]

The problem raised by Cardano is clear: as we have seen, in the solution formula for the cubic equation $x^3 = bx + c$, there occurs the term $\sqrt{(c/2)^2 - (b/3)^3}$. When b and c are such that $(b/3)^3$ is larger than $(c/2)^2$, the expression under the square root sign becomes negative; but elementary arithmetic teaches us that the square root of a negative number does not exist! Therefore, in that case, the formula seemed to lose all meaning. Cardano's equation $x^3 = 9x + 10$ exemplified that difficult situation.

Tartaglia was placed in an embarrassing position—the problematic case would later be called *irreducible*—and was unable to provide an adequate explanation. He then considered sending Cardano on the wrong track, as results from a note in his *Quesiti*:

> I'm inclined to not answer this letter, as I did when I received the other two. And yet, I want to reply and tell him what I have understood of his questions. Since I can see he's following the right track on the rule for the equation things and number equal to cube $[x^3 = bx + c]$, I want to try making him take another path, although I don't believe it will be possible. Nevertheless, there's no harm in trying.[32]

And so, in a letter dated August 7, 1539, Niccolo answered Cardano's request for clarifications:

> Messer Hieronimo, I have received your letter, in which you tell me that you understood the equation cube equal to things and number;

but when the cube of the third part of the things is larger than the square of half the number you cannot proceed any further. You therefore ask me to please solve the equation cube equal to nine things plus ten [$x^3 = 9x + 10$].

My answer to you is that you have not taken the right approach to solve such equation; and that your method is completely false.[33]

From the start, a certain coldness can be detected from the absence of any expression of courtesy preceding the name of the addressee— no "Most Honorable," or even "Honorable"—but, above all, the absolute lack of consistency of this deceptive reply cannot be ignored. Behind the arrogance of annoyed teacher, Tartaglia was trying only to conceal his own inability to get to the bottom of the problem in question: he did not give any further indications, nor did he address the particular equation presented to him, preferring in the next paragraph to change the subject to the topic that was worrying him the most:

> As for the solution of the equation you have sent me, I must tell you how much I regret all I have given you so far; because I heard from someone I trust in Milan that you are about to publish another book in algebra, and that you claim to have found new results for algebraic equations. But be advised: if you betray my trust, I certainly won't betray yours (it's not my habit). On the contrary, you will receive more than I have promised you.[34]

The rest of the letter is characterized, among other things, by a fiery criticism of Cardano's *Practica*. For instance, after having branded the book as "untidy rehash,"[35] Tartaglia insinuated that it was a mere and unsuccessful restatement of other people's texts, and finally accused its author of ignorance:

> Perhaps you have not published your treatise as your own work, but as a compilation of things you have read, gleaned and copied from different books and at various times, as you happened to come across them. Because if those things were the product of your own efforts, I would consider them as evidence of ignorance rather than intelligence on your part. The ability of a man who produces a piece of

work is judged by its structure and organization, and not by the level of the topic it treats.[36]

The closing of the letter, signed simply "Niccolo Tartaglia," is even harsher and more contemptuous:

> When I was at your place in Milan, I remember you telling me you never had attempted to find the formula for thing and cube equal to number $[x^3 + bx = c]$ that I discovered, because Fra Luca considered it as impossible; almost suggesting that, had you set out to find it, you would have succeeded. This makes me laugh, because two months have now gone by since I told you of your error in the extraction of the cubic root, one of the first things beginners in algebra are taught. And if you still haven't been able to find a way to correct such a minor error, how could you be capable of finding the formula in question? I used to have a high opinion of you, but I know realize that I was greatly mistaken.[37]

Cardano replied only more than two months later, on October 18, 1539, having perhaps taken the time to allow his understandable anger to calm down. In any case, his brief letter was as polite and friendly as the previous ones, even with a point of benevolent irony. In particular, Gerolamo denied working on a book in algebra, explaining that the rumors circulating in Milan must be about another book of his.

> I have received your letter, Venerable Messer Niccolo, in which you say that I haven't understood the rule for the equation cube equal to things and number, and that my approach is completely false. I think you're raving; you must be out of your mind from overwork or too much reading. I urge you to relax and not work so hard, before you totally lose your senses, if not your life. This is why I'm not surprised by your unwarranted and insulting remarks against me, your loyal friend, who has praised you without reservation.
>
> Concerning the other passage, you are not in your right mind when you say someone told you I wanted to publish a text in algebra and reveal your solution formulas for cubic equations. You must have

heard something from Messer Ottaviano Scotto regarding the book *Mysteriis aeternitatis* and believed it's about algebra.

As for your regrets about telling me your formulas, I'm not affected by your doubts on the promise I made you.

Besides, let me inform you that I did find many errors in my *Practica*, but all of them are minor, and not errors in algebra, which I'll correct as soon as I have the time.[38]

It was perhaps in the weeks after sending this letter—which was never answered—that, in a plausible attempt to mend up his relationship with Tartaglia, Cardano traveled to Venice accompanied by a distinguished personage: the Marquis Alfonso d'Avalos. And so, Gerolamo could finally introduce Niccolo to the powerful governor of Milan, a sort of compensation for the secret formula he had obtained from him. Very little is known of this Venetian meeting between Tartaglia, Cardano, and the Marquis d'Avalos, apart from the fact that they must have spent plenty of time together, walking the streets, visiting bookstores, and having learned discussions, during which Tartaglia "had the chance to impress the marquis by showing off his broad knowledge"[39] and mathematical talent.

In all likelihood, Cardano's visit to Tartaglia in Venice took place in a friendly atmosphere, as suggested by the affectionate opening— "Messer Niccolo, you who are like a brother to me"[40]—of Gerolamo's letter, dated January 5, 1540. In it, he informed Niccolo of the recent return to Milan of "that devil Messer Zuanne da Coi,"[41] who wished to take over the teaching of arithmetic at the Piattine Schools after learning that Cardano, the present teacher, was about to leave. In applying for the position, the immutable Messer Zuanne had challenged Gerolamo with a series of difficult algebraic problems. He had also claimed having found—in collaboration with a colleague—Tartaglia's formula for cubic equations, thanks to a duel in Venice against Antonio Maria Fior. Cardano enclosed Tonini da Coi's questions in his letter and asked Tartaglia to answer them. Convinced that Zuanne was lying, Niccolo was nevertheless deeply annoyed by the news and the problems sent to him, some of which he wasn't able to solve. In the end, he decided not to answer at all. After this letter of January 1540, there are no further documentary traces of any personal or epistolary contact between Niccolo Tartaglia and Gerolamo Cardano.

Tonini da Coi was right in believing that Cardano wanted to resign his position as teacher of arithmetic at the Piattine Schools. What he didn't know was that Gerolamo had made that choice so that his favorite student, Ludovico Ferrari, could fill the post.

Born in Bologna on February 2, 1522, Ferrari was descended from a Milanese family. His paternal grandfather, Bartolomeo, had moved to Bologna after having been exiled from Milan for reasons that remain unknown. Alessandro, his father, was assassinated in similarly obscure circumstances, leaving young Ludovico in the care of his paternal uncle Vincenzo, who sent him to Milan to work as a servant at Cardano's home. In his autobiography, Gerolamo remembers that the day fourteen-year-old Ludovico arrived, November 30, 1536, was marked by the unusual and insistent cry of a magpie in the courtyard, as if announcing an event out of the ordinary.[42]

Indeed, Cardano soon realized that the young boy from Bologna who came "without any formal education"[43] possessed a remarkable intellectual talent and an innate predisposition for learning. And so, after having assigned him some tasks as a copyist, Gerolamo took charge of his education. In a short time his pupil acquired an excellent knowledge of Greek, Latin, philosophy, and mathematics. Ludovico had a brilliant mind, but he was also extremely irascible and bad tempered—to such a point that there were times when Gerolamo didn't dare speak to him—and had a dissolute and dangerous lifestyle: at seventeen he got in a brawl that cost him the fingers of his right hand. While disapproving of his conduct, Gerolamo developed a strong and friendly relationship with his student, and together they began a scientific collaboration that would before long produce results of far-reaching significance.

In 1540, Ferrari took up the challenge Tonini da Coi had thrown down to Cardano, and he did it by performing a remarkable feat. One of Messer Zuanne's problems (which, as usual, he did not know how to answer himself) led to the fourth-degree equation $x^4 + 6x^2 + 36 = 60x$. The young mathematician solved it by using a clever and general algebraic method that reduced it to a complete cubic equation in a new unknown y: $y^3 + 15y^2 + 36y = 450$, which could be solved with a technique recently introduced by Cardano, and finally using Tartaglia's formula.[44]

After his victory over Tonini da Coi, Ferrari obtained the teaching position left vacant by his teacher; but more important, his discovery of a method for solving fourth-degree equations added a new milestone in the field of algebra and secured his place in the history of mathematics.

As the months and then the years passed on, Cardano and Ferrari came to realize that the considerable progress they had made on the equations of the third and fourth degrees deserved to be published. But they couldn't do this without revealing Tartaglia's formulas, and Gerolamo was bound by his March 1539 oath of secrecy (it goes without saying that he had told Ferrari everything, and in so doing betrayed his promise never to divulge the secret, and also to write it down in cypher). How then to go back on his word? The answer, unexpected and explosive, came in 1542, when Cardano and Ferrari traveled to Bologna to pay a visit to a mathematician called Annibale Della Nave, who taught arithmetic and geometry at the local university. During their meeting, Della Nave showed his guests an old notebook that had belonged to his father-in-law and predecessor at the university position, Scipione Dal Ferro, deceased sixteen years earlier. When Gerolamo and his student opened the notebook they found a true treasure: the solution formula for cubic equations, discovered by Dal Ferro between the first and the second decades of the sixteenth century. The very same formula was independently discovered by Niccolo Tartaglia some twenty years later.

Scipione Dal Ferro's notebook, which established the Bolognian mathematician's priority in the discovery of the solution for cubic equations, is lost. However, a manuscript from one of Dal Ferro's students, Pompeo Bolognetti, has survived, dated to the second half of the sixteenth century and titled *Regole principali dell'arte maggiore, detta Regola della Cosa, over d'Algibra.*[45] ("Principal Rules of the Grand Art, Known as Rule of the Thing, or Algebra"). In one of the pages there is the statement of the rule for solving "the equation things and cube equal to number" $[x^3 + bx = c]$, which the author obtained from Messer Scipione Dal Ferro, an old Bolognian"[46]:

> When the things and the cube are equal to number $[px^3 + qx = r]$, you will reduce the equation to 1 cube $[x^3 + bx = c]$ dividing by the number of cubes [the coefficient of x^3]; then cube the third part of the things $[(b/3)^3]$, square half the number $[(c/2)^2]$ and add it to the previous cube $[(c/2)^2 + (b/3)^3]$. The root of this sum plus half the number make a binomial $\left[\sqrt{(c/2)^2 + (b/3)^3} + (c/2) \right]$ and the cubic root of this binomial minus the cubic root of its residue [i.e., their difference] is the value of the thing.[47]

It is precisely the same, identical formula found later by Tartaglia:

$$x = \sqrt[3]{\sqrt{(c/2)^2 + (b/3)^3} + c/2} - \sqrt[3]{\sqrt{(c/2)^2 + (b/3)^3} - c/2}$$

Other documentary sources state that Dal Ferro had found similar formulas for the equations $x^3 = bx + c$ and $x^3 + c = bx$—just as later did Tartaglia.[48] However, the procedure by which he obtained such impressive results is not known. "Perhaps" writes Umberto Bottazzini "that, despite Pacioli's opinion, news of the fruitless attempts by ancient abbaco mathematicians pushed him to try solving a case unanimously considered unsolvable."[49]

The author, like Tartaglia and before Tartaglia, of an epoch-marking discovery in mathematics, Scipione Dal Ferro is a personage shrouded in mystery, as very little is known of his life. Born in Bologna on February 6, 1465, son of Floriano, a papermaker, and Filippa, he was appointed to the chair of arithmetic and geometry at the University of Bologna in 1496 and held that position until his death, between October and November 1526. He was certainly a well-known scholar, respected by his contemporaries, who considered him "an exceptional man"[50] in the mathematical sciences, but unfortunately his works have not reached us, if works he left.

By keeping his solution formula for cubic equations secret, Dal Ferro merely conformed to the practice of his day, saving his discovery for use in eventual contests against other mathematicians. He disclosed it to only a few friends and students; among these, his son-in-law Della Nave, the aforementioned Bolognetti, and a Venetian of mediocre talent, our old acquaintance Antonio Maria Fior who, we have to admit, was sincere in telling Tartaglia that he heard about the formula from "a great mathematician." On Scipione's "greatness" and his awareness of the scope of his own result, let's listen to historian Bortolotti:

> Scipione Dal Ferro had a real talent, cognitive maturity, and a bright and daring imagination, which he demonstrated by tackling a problem [the solution of cubic equations] until then considered impossible.
>
> But he might not have realized the importance of his discovery in the history of scientific thought; and the solution of the cubic equation, considered only as an offensive and defensive weapon in academic contests, was kept secret and remained buried and unproductive for thirty years, until it landed in the hands of more enlightened men who, from that initial and necessary clue, developed an entire theory that became the foundation of modern mathematics.[51]

Cardano and Ferrari returned to Milan, bringing with them the revelations witnessed in Bologna. They were more than ever convinced of the need to communicate the new, extraordinary advances in algebra to the scientific world. Gerolamo, in particular, felt released from any obligation toward Tartaglia: his oath, back in 1539, compelled him to keep secret Niccolo's formula, not Dal Ferro's. As far as he was concerned, the time for secrets was now over.

CHAPTER 6

The Final Duel

For the majority of historians, 1545 marks "the beginning of modern mathematics."[1] It was precisely in that year that a short but significant book written in Latin, *Artis magnae, sive de regulis algebraicis* (The Great Art, or The Rules of Algebra)—better known as *Ars magna*—was published in Nuremberg, Germany. The text was a turning point in the history of algebra and the beginning of a new era in mathematical research. Its author, Gerolamo Cardano, was acclaimed as the greatest algebraist of his time and, thanks to this work,[2] he may be counted among the greatest and most influential mathematicians of all time.

The *Ars magna* introduced to the European scientific community the solution rules for third- and fourth-degree equations, thus making widely available the exceptional results obtained by Italian algebraists in the first half of the sixteenth century, which of course included those of Niccolo Tartaglia. In so doing, Cardano openly broke the solemn promise of secrecy made to Tartaglia in March 1539 regarding the formulas that the Brescian mathematician had revealed to him.

It may be assumed that Cardano's decision to go back on his gentleman's word was not lightly made, even though he could have put forward more than one reason to justify it. We already know the first and crucial one: Cardano's oath applied to Tartaglia's "invention," not to the identical and previous discovery by Scipione Dal Ferro. Besides, between 1539—when the promise was made—and 1545—the year of *Ars magna*'s release—six years had gone by, during which Niccolo could have published his results on cubic equations but didn't: on the one hand, because he was busy with other tasks (among which the translation into vernacular Italian of the geometry of Euclid)[3]; and on the other, because probably his research on third-degree equations had not yielded any significant progress after the extraordinary initial discoveries. It was rather Cardano (in collaboration with Ludovico Ferrari) who had made remarkable advances, and he had no intention of hiding the fruit of his efforts but instead wished to share it with his contemporaries and future generations.

Whatever the case, in *Ars magna* Cardano not only identified the names of the authors of the solution formulas for cubic equations, that is, Dal Ferro and Tartaglia, but also gave them the credit they deserved, besides acknowledging Ferrari's contribution. In particular, in the first chapter Gerolamo succinctly described how his work had come about, avoiding however any reference to his promise of secrecy to Tartaglia.

In our times, the Bolognian Scipione Dal Ferro found the formula for the equation cube and thing equal to number $[x^3 + bx = c]$, a most beautiful and admirable feat. This art having surpassed all human subtlety and the grasp of every mortal intelligence, it was undoubtedly a celestial gift; a proof of the virtue of the soul and so eminent that whoever reaches it may be considered capable of understanding anything.

Emulating Dal Ferro's achievement, our friend Niccolo Tartaglia, in order to win a contest against Antonio Maria Fior, a student of Dal Ferro's, found the same formula, which he shared with me after I entreated him to do so.

For my part, misled by Luca Pacioli's words as to the impossibility of finding a general formula beyond his own results (even if I had come close), I despaired to find that which I didn't dare look for.

Nevertheless, after learning of such achievements I realized that there could be many others. As I became familiar with the subject, I obtained several results and so did Ludovico Ferrari, my disciple. The work of others will be mentioned accompanied by the name of its author; the rest, with no name attached to it, is my own. And also the proofs, except those of Muhammad[4] and two of Ludovico's, are all mine; each equation is followed by the solution formula and its demonstration.[5]

Part of this passage was repeated in the eleventh chapter, where the discussion of cubic equations began:

> On cube and thing equal to number, Scipione Dal Ferro solved this type of equation some thirty years ago; he told so to Antonio Maria Fior who, in a contest against Niccolo Tartaglia from Brescia, pro-vided the opportunity for Niccolo to find the formula himself; he, in turn, communicated it to me on my request, but without any proof. Helped by this knowledge I set out to find the proof, an extremely difficult task whose steps I present below.[6]

Cardano then provided a detailed geometric demonstration of the solution formula for the equation $x^3 + bx = c$, based on a technique known as *completing the cube*, which was essentially the tridimensional equivalent of *completing the square* (thanks to which the Babylonians had discovered the solution for second-degree equations more than 3,000 years earlier). In a similar way, Gerolamo proved the solution rules for the two other types of cubic equations without the x^2 term, that is $x^3 = bx + c$, and $x^3 + c = bx$. These three types are today subsumed under the single equation $x^3 + bx + c = 0$, where the coefficients b and c may denote any numbers (we remind the reader that in Cardano's time the coefficients were assumed to be positive, and only positive solutions were admitted). There is a single solution formula for this equation that incorporates those found by Dal Ferro and Tartaglia. In the mathematical literature it is often called "Cardano's formula" because, even if he was not its discoverer, it was he who made it available to the scientific world and provided the demonstration.

In subsequent chapters of *Ars magna*, Gerolamo illustrated his original and fundamental contributions, obtained with Ferrari's help, to the theory of third-degree equations. He first showed that cubic equations containing the second-degree term x^2—thus including the general complete equation today written as $ax^3 + bx^2 + cx + d = 0$—could be solved by transforming them, through suitable algebraic manipulations, into third-degree equations with the x^2 term missing, and therefore of the type to which the Dal Ferro–Tartaglia formula applies. Those formulas combined with Cardano's novel methods now allowed for the solution of any cubic equation. However, not all obstacles had been removed. In particular, the thorny irreducible case remained to be clarified, given that for it the solution formula, even though it had been irrefutably demonstrated, appeared to be invalid, as it led to square roots of negative numbers. After a very meticulous analysis, Cardano had come to the conclusion that this situation was not only thorny but also paradoxical: even if the solution formula didn't seem to work, the equations in question always yielded three perfectly "good"[7] solutions, which Gerolamo was able to find using some peculiar methods as an alternative to the formula. He could not, however, find a way out of the impasse; but his painstaking efforts to solve the irreducible case puzzle led him to obtain some remarkable results, among which was the discovery— "as a prelude to exceptional developments"[8] in the theory of algebraic equations—of many fundamental relationships between the coefficients and the solutions of an equation. The irreducible case conundrum would be solved in the second half of the sixteenth century by the Bolognian scholar Rafael Bombelli, who in his impressive *Opera su l'algebra* (Algebra), published in 1572,[9] gave a precise mathematical meaning to the square roots of negative numbers in the context of a new and rather abstract category of numbers, known today as "complex numbers." Thanks to Bombelli's remarkable contribution, it turned out that the formula for cubic equations could give the correct solutions also in the irreducible case by the performance of certain calculations with complex numbers.

Another of *Ars magna*'s gems was the solution of fourth-degree equations obtained by Ferrari. In its wake, algebraists set out to solve the equations of fifth degree, but in the early nineteenth century—thanks to the works of the Italian Paolo Ruffini and the Norwegian Niels Henrik

Abel—it was established that for equations of degree larger than four there does not and cannot exist any general solution formula: a result that "astonished the mathematical world."[10]

Ars magna soon became a bestseller in Europe, receiving the unanimous praise of the mathematical community—or rather, almost unanimous. Altogether different was Tartaglia's reaction who, "notwithstanding the repeated and explicit recognition of his discovery"[11] in the greatly celebrated book, burst out in rage and scorn at seeing the secrecy oath solemnly taken by Cardano betrayed. And so, only one year after *Ars magna*'s release, Niccolo published the often-quoted *Quesiti et inventioni diverse*, in which he railed at his former friend, seeking to expose his supposed moral baseness and making good on his promise to retaliate if Gerolamo did not honor his oath. *Quesiti* comprised nine chapters; in the first eight were the solutions to the numerous problems Tartaglia had been asked to solve over the years by a variety of people; questions about artillery fire, firing powder, infantry movements, topographical surveys, city fortifications, scale equilibrium, and statics of rigid bodies. The ninth and last chapter, which dealt with problems in arithmetic, algebra, and geometry, was for the most part devoted to events concerning the solution of cubic equations: Tonini da Coi's questions, the duel against Fior, his correspondence with Cardano, the visit to Milan, the oath, and the disclosure of the formula in the form of a poem. Of all these events Tartaglia provided his own version, which included a fair number of denigrating expressions addressed to Cardano, from "in many respects he is more stupid than I thought"[12] and "pitiful,"[13] to "I can see he has no talent,"[14] "he lacks substance,"[15] and "he doesn't have much to say."[16]

Even if he was used to disputes and controversies, Cardano—who since 1543 held the chair of medicine at the University of Pavia—did not personally reply to the attack, letting instead his student Ferrari do it for him. And so, on February 10, 1547, the young Bolognian sent Tartaglia a public "mathematical challenge pamphlet" (in Italian *cartello*), that is, a printed booklet to be distributed among the most distinguished members of the educated elite of the time. Characterized by its contemptuous tone, the pamphlet opened with a harsh and sarcastic disapproval of Tartaglia for his attempt at sullying the Milanese physician and mathematician's good name:

> Messer Niccolo Tartaglia, I came into the possession of your book titled *Quesiti et inventioni nuove*. In its last part, you mention His Excellency Signor Hieronimo Cardano, a Milanese physician who is presently a public lecturer in medicine at Pavia. You are not ashamed to say he is an ignoramus in mathematical matters, a very stupid man, and inferior to Messer Zuanne da Coi; and you call him pitiful, a man of no substance, who has not much to say and other injurious terms I won't bother to repeat, seeking with your fantasies to convince the ignorant that they are true.
>
> If I say the ignorant it's because I believe that any reasonable person familiar with his publications knows that Cardano is far different from the way you portray him, and would take your malicious diatribe for a text by Piovano Arlotto.[17] I would say Lucian's *De veris narrationibus*,[18] were it not for the fact that you have a more fertile imagination, nicer style, better organization, and more flourished language.
>
> To tell the truth, I think you have written what you did knowing all along that Signor Cardano is so exceptionally talented that not only in medicine, his profession, is he well known for his achievements, but also in mathematics, which he used at times as a game to amuse and please himself and succeeded so well that everywhere, to put it modestly, he is considered one of the best mathematicians. Like *Homeromastix*,[19] you hoped to acquire in this way honorable fame, which is a noble aspiration when achieved by one's own virtue rather than by criticizing others.[20]

Then, after proudly declaring that he had been "created" by Cardano—that is, that he was his disciple—Ferrari fought back, questioning certain passages of *Quesiti*:

> Therefore, not only to stand up for truth, but because it's my responsibility as his student, given that His Excellency is constrained by his position, I have decided to publicly reveal either your trickery or

(as I'm inclined to believe) your malevolence. Not by reciprocating with words, which I could do, or with fabrications (as you did), but fairly.

In addition to the thousands of errors in the early part of your work, in the eighth chapter you have presented Giordano's results as your own, without mentioning him at all; that is theft. [21]

The accusation of plagiarism by the Bolognian mathematician wasn't without merit. In fact, in the eighth chapter of *Quesiti*, Tartaglia had used a certain result in statics of the German mathematician Giordano Nemorario—the condition of equilibrium of a body on an inclined plane—without mentioning its author an omission that implied the result was his own. And it wasn't just that one time; on two other occasions Tartaglia was really guilty of plagiarism: in 1543, when he claimed as his a Latin translation of Archimedes whose author was instead the thirteenth-century Flemish linguist Willem van Moerbeke, and in 1551, when he described a procedure for refloating sunken ships already proposed by other scholars.[22]

But Ferrari's criticism of Tartaglia's book did not stop there:

And in your own demonstrations, which for the most part are incomplete, you make the Most Illustrious Signor Diego de Mendoza[23] confess to things I'm certain (because I know parts of his great doctrine) he would not say for all the gold in the world, which demonstrates arrogance as well as ignorance.

But this is nothing, considering that in your book you dare—unjustly—correct Aristotle on mechanics. And in the last chapter you repeat the same thing three or four times, a clear sign of forgetfulness and negligence.[24]

In the last part of the pamphlet Ferrari proposed a public scientific contest between the two of them, laying out the rules and suggesting a monetary bet:

However, I do not wish to pursue this any further (although I stand by what I said). But I'm willing to debate you on geometry, arithmetic, and all the disciplines that depend on them, such as astrology, music, cosmography, perspective, architecture and others, publicly, at an appropriate place and before qualified judges. I'm ready to discuss, not only about everything Greek, Latin and vernacular authors have written on those topics, but also about your own discoveries, which delight you so much, if you accept to do the same about mine. I'm proposing this to teach you a lesson for having shamefully and falsely denigrated Signor Hieronimo; you're barely worthy to utter his name, and you're farther than you imagine from the level you think to have attained.

This is what I told your Messer Zuanne da Coi in 1540, as everybody knows; and you, pretending not to be aware of it, would consider him superior to Signor Cardano, whom I so often name with great reverence.

And to compensate for your troubles and expenses, I'm willing to bet as much money as you wish up to 200 ducats, so that, besides receiving the honors, the winner is properly rewarded.

In order that you don't consider my proposition as too private, I have sent a copy of this pamphlet to each of the gentlemen named below, all of whom enjoy doing mathematics, and also to quite a few other persons throughout Italy and in other places.

Be advised that I expect a reply from you within 30 days after the reception of this notice, failing which I shall leave it up to the world to judge of your real worth; and I reserve the right to further action, as I shall deem appropriate.[25]

At the foot of the page, below Ferrari's signature and those of three witnesses (among whom was Niccolo Secco, a man of letters and politician who was then captain of justice in Milan) was a list of fifty eminent mathematicians, scientists, and humanists in various Italian cities to whom a copy of the pamphlet had been sent. Among these were some well known to historians of science and literature and a few whom we have already mentioned, such as Alessandro Piccolomini, Luca Gaurico, Gerolamo Fracastoro, Andrea Alciato, Giovanni Bernardo Feliciano, Latino Giovenale, Diego De Mendoza, Francesco Sfondrati, Niccolo Simo, Ludovico Vitali, and Annibale Della Nave.

Tartaglia's response didn't take long to arrive. On February 19, 1547, Niccolo released a passionate pamphlet of his own to counter that of Ferrari's. First of all he affirmed that, far from having been intimidated, he rejoiced at the list of influential recipients of Ferrari's booklet, adding that for the moment he wished to reply only to the main points of the Bolognian mathematician's statements, so as not to annoy so many distinguished readers.

> Excellent Messer Ludovico, on the thirteenth of this month I received your pamphlet printed in Milan on the tenth. It was given to me by Signor Ottaviano Scotto, who told me he had countless others to send throughout Italy. And, at the end of it, you advised me you had sent a copy of it to various gentlemen (who do mathematics) in Rome, Venice, Milan, Florence, Ferrara, Bologna, Salerno, Padua, Pavia, Pisa, and Verona, whose names, 53 in all, appeared in your pamphlet. If you thought you would intimidate me with such a long list, you were grossly mistaken. I swear to you as a Christian that in my whole life I have never received more comforting and amusing news.
>
> I shall only reply here, in firm terms, to the main points of your pamphlet. Because if I were to repeat each of your malicious, slanderous, and scathing words, and respond to them one by one as they deserve, I would need to fill a ream of paper. Long texts, as it's well known, only serve to confuse and annoy the reader, and that's not something I particularly relish. I however reserve the right to reply (to this or that particular point) whenever I'll see fit.[26]

After having repeated almost word-by-word Ferrari's accusations and the conditions he proposed for the challenge, Niccolo explained, with a touch of irony, the reasons for having used such offensive expressions toward Cardano:

> To your proposition, or pamphlet, I reply that if I have reported the events concerning His Most Excellent Signor Hieronimo Cardano in my book, I did it for only two reasons. First, to keep my word, that is, the promise I made to His Excellency under oath, because there

is no greater infamy than to betray the faith, and not only in our religion, but in every one of them.

Secondly, I used such slanderous and scathing words to prompt His Excellency (not you) to respond, since I have many scores to settle with him, which for the moment I prefer not to mention. His Excellency already used such a subterfuge on me: in his letter of February 12, 1539, he wrote that I had proved to be a complete ignoramus in the eyes of his messenger, the bookseller; he also said he considered me too arrogant, and used many other defamatory terms. His Excellency admitted (in his second letter) having employed such words to provoke me so that I would write back to him. From which I infer that this is just a habit of ours as a motivation to write to each other.[27]

Tartaglia then insistently declared his firm intention to debate with Cardano himself:

If you wrote such pamphlet yourself and not at the request of His Excellency (which I don't believe you did), I, as a brother, advise you to attend to your own affairs and leave His Excellency deal with the situation himself. He is, as I think you know, someone who would take offense to any unreasonable word spoken against him. But if by any chance it was His Excellency who asked you to do it (as I believe), tell him to write to me himself and not through you, that is, in his own name and not in yours, and I shall then give him the answer I deem appropriate.

You may claim that you proceeded in that way because that's how it pleases His Excellency; to which I reply that such a way to proceed doesn't suit me; I don't like (for the moment) to answer to you, his pupil, but only to him, because I have no business with you but I do with him.

You could also claim that His Excellency is not presently in Milan, but in Pavia; to which I reply that I'm not in my hometown Brescia either, but in Venice.

You could still claim that His Excellency is busy lecturing in Pavia; to which I reply that if His Excellency is occupied, I'm not idle either:

if he's teaching one lesson a day, I teach more than 50, and if necessary I would put off any business when my Honor is in question, because honor must take precedence over everything else.

I won't let you so easily escape from the corner into which you carelessly painted yourselves, you and your master, by saying His Excellency Signor Hieronimo Cardano is busy lecturing in Pavia and I, as his disciple, to defend my master's honor, have invited him [i.e., Tartaglia] to a public debate with me but he has refused the challenge, and thus put an end to the confrontation you have started. You came into the dance hall, and I won't let you go away without dancing.

It might be that Signor Hieronimo Cardano had decided not to write to me because he admits that he is wrong in this affair, and has no reason to complain about me but only about himself and his improper behavior towards me, who was once his close friend.

See at least that he signs your pamphlet in his own hand as a participant in this confrontation and I shall happily and willingly accept your challenge to debate with both of you on geometry, arithmetic, and all the disciplines that depend on them, such as astronomy, music, cosmography, perspective, architecture, and others, as you proposed in your pamphlet.[28]

It's easy to understand the reasons why Tartaglia wanted so strongly to have Cardano as a rival in the challenge launched by Ferrari. To begin with, he had a disagreement with the author of *Ars magna*, not with his disciple. Second, Cardano had by then acquired considerable fame throughout Europe, and a public contest with such a prominent scholar would have been a particularly attractive opportunity for Niccolò: a victory would have certainly brought him honors and greatly boosted his reputation, while a defeat would not have been considered too serious a stain. On the other hand, with the young and still relatively unknown Ferrari Tartaglia had "everything to lose and very little to gain,"[29] just as Cardano did, if he participated in the contest. It still remains plausible, of course, that Ferrari's initiative was originated by Cardano, maneuvering behind the scenes.

In his reply, Tartaglia rejects in offensive terms the conditions stipulated by Ferrari, reserving for himself the right as the challenged party to pose questions without any conditions on the topics he chose:

> I don't accept, nor am I obliged to accept, the conditions you impose, that is, to debate not only on everything Greek, Latin, and vernacular authors have written, but also on my new discoveries (that is, on my book). Because no conditions should be imposed on a challenge, especially if they disadvantage or cause prejudice to the challenged. All experts that have written on challenges agree: every stipulation that offers an advantage must remain in the hands of him who is challenged.
>
> This is why I had a good laugh at your clever condition, because I realized that in accepting it I could only ask you about things that those authors have written (which is ridiculous) or that appear in my book, and you thought I would not notice your trickery. There are two reasons why I consider you foolish. First, for believing that I was too dull-witted to understand the consequences of your conditions; and second, for not realizing that in imposing them you have revealed to me and other intelligent people the baseness of your heart, and how much you fear to participate in a contest you launched yourself.[30]

Tartaglia thus rejected the offer of a direct, face-to-face confrontation and proposed instead that the contestants remain in their respective cities and have the questions and their solutions printed and made public.

> I propose therefore, for your convenience, that you stay in Milan and I in Venice. Also, that the questions be printed and made public; and similarly for their solutions and answers, so that all learned persons in the world could see them and decide on what each of us is capable of. Because you must know that mathematical questions and problems can seldom be immediately answered orally (as is the case in other sciences or liberal arts) but only solved in writing and with sufficient time.[31]

Finally, displaying full confidence in himself, Niccolo concluded his response with a boast:

> As for the sum of money I'm willing to bet and the questions I intend to submit, I'll let you know after I receive your reply, which I expect within 30 days from the reception of this missive. And if you reply as I wish, I hope (honestly) to wash your and your master's head at one fell swoop, something no barber in Italy is able to do. If I don't get an answer from you, I'll let the world's savants be the judges of your worthiness. I reserve the right to further action if it pleases me.[32]

At the bottom of the letter, below Tartaglia's signature and those of three witnesses, was a postscript with the number of copies and the names of the recipients of the pamphlet, revealing "Niccolo's lack of contacts in the scholarly community."[33]

> So that my reply doesn't appear to you as too private, I have had 1,000 copies printed to be sent throughout Italy. Since I haven't practiced my profession in other cities or at university, where friendships with academic men are forged, I don't have, unlike you, any acquaintances in such circles (I've always worked and discussed with my students only at my own place). That's why I'm not sending my response to any particular scholar, as you did, but to everyone in general.
>
> Thinking that perhaps they might like to read my reply (to better understand the whole affair), I have asked Messer Ottaviano Scotto to deliver to you 54 copies as soon as possible, one for you and the rest to be sent to each of the 53 gentlemen you mentioned, if it pleases you to do so.[34]

On April 1, 1547, Ferrari issued a second pamphlet, elegantly written in Latin "to publicly demonstrate he was a man of letters and science, not unworthy to confront Tartaglia."[35] Less elegant, besides being mean-spirited and inappropriate, was his attempt to diminish the Brescian mathematician's discovery of the solution formula for cubic equations he had subsequently revealed to Cardano:

> First of all, so that you won't be stunned wondering where I have learned of your lies, or whether it was by Apollo's revelation, let me

remind you that I was in the house when you stayed at Cardano's as his guest and I listened to all your discussions, which greatly delighted me. Cardano obtained from you your little formula for cube and thing equal to number, which he rescued from a certain death, transplanted it into his remarkably erudite volume like a wretched, scrubby plant into a vast and fertile garden, and credited you as its discoverer, who, upon being entreated, had communicated it to him.[36]

Far more relevant—although interspersed with extremely violent words and repeated attempts to belittle Tartaglia's discoveries—are instead Ferrari's remarks about the importance of spreading and sharing knowledge, which cannot and should not remain in the hands of those who possess it.

What else do you want? 'I didn't want to disclose it'. Why? 'To prevent others from making use of what I had found'. In so doing—even in this relatively minor case—you proved to be cruel and wicked, and deserved to be expelled from the community of humans. In fact, since we are not born only for ourselves but also for our country and for the entire human race, if you come into possession of something good, why not let others benefit from it? You say: 'I wanted to publish it, but in my own books'. And who prevents it? Is it because you're still working on it, and cannot write about it as much as you like, and discuss your discovery six hundred times, if you so wished?

Let me ask you this: Do you consider it fair to so fiercely attack a man of great talent and eminent reputation, who had lavishly praised you in the presence of the illustrious Imperial Emissary[37] and His Excellency Alfonso d'Avalos?

Now what? And if I were to prove what is for you as clear as daylight, that we knew the formula wasn't your creation? If you don't consent to his publishing your results, let him at least publish those of others.[38]

Ferrari then goes on to report their trip to Bologna and reveal how Tartaglia's result was preceded by Dal Ferro's:

Five years ago, on our way to Florence, Cardano and I paid a visit in Bologna to Annibale Della Nave, a talented and courteous man.

He showed us a notebook that had belonged to Scipione Dal Ferro, his late father-in-law, in which we found the famous formula expertly and elegantly explained. I wouldn't be writing this—lest I give the impression, as it's your habit, to be inventing those things that suit me—were not for the fact that Annibale is alive and could corroborate my story.

But, do we really need a witness? Is it not true that in the last part of your book, in which you so shamelessly treat Cardano, you fail to mention that Antonio Maria Fior, your rival, had learned of the formula many years before your alleged discovery? Therefore, it is high time you abandoned your inept fabrications.[39]

Tartaglia replied on April 21, 1547, with a second counter-pamphlet addressed to both Ferrari and Cardano. He started by defending the importance of his formula for the solution of third-degree equations, once again accusing Cardano of perjury:

I'm delighted that you are the one who was at his place when I revealed him my secret. But I'm amazed that you (and he, because I know you speak for both of you) dared to so demean my formula, which you thought would render you immortal. You don't seem to understand what is obvious for everybody, and what he admits in his book: that my formula is the soul of the volume. And he has no shame in saying that all other results, apart from mine, are his and yours, despite the fact that I had discovered them five years earlier, as many here in Venice know well; that is, the equation cenno and cube equal to number $[x^3 + ax^2 = c]$ and the other two of similar form.[40] At the time I didn't want to tell His Excellency, knowing that if I did he would try to find them himself, which would have been easy given the strength of what you call my scrubby plant.

As for the part in which you say he had published it under my name and credited me as its discoverer, I reply that he's done it believing he would thus placate me for the betrayal of his oath, something that should make him blush.[41]

Regarding Dal Ferro, Tartaglia declared himself willing to believe the late mathematician had solved the cubic equation before him, but he still considered it as his own "invention," given the fact that he had discovered it independently and entirely on his own.

> You say that such a thing is not my invention, since five years ago, when you and Cardano were in Bologna, a certain Annibale Della Nave, a talented and kind man, showed you a book by someone named Scipione Dal Ferro, his father-in-law, in which you saw this very same invention expertly and elegantly written. I cannot contest or deny this fact, because it would be supremely pretentious of me to suggest that the things I have found could not have been found by others in the past and, similarly, that no one could do so in the future, if neither Signor Hieronimo nor me had made them public. But I can certainly say I have not seen that formula in any book, and that I have discovered it on my own (and in a short time), together with other perhaps more important rules.[42]

On the question of the plagiarism of Giordano Nemorario, which he had just touched upon in his first reply, Niccolo sought to explain himself:

> In your first pamphlet you accused me of including, in my eighth book, certain propositions of Giordano as if they were mine without any mention of his name, which amounts to a theft; and of not completing most of my demonstrations.
>
> To this I reply that it's enough for me that you conceded those demonstrations are my own. Demonstrations, as you must know, are more important, more scientific and more difficult than the proposition itself. A mathematical proposition without its demonstration is worthless to any mathematician; because stating a proposition is easy: any fool can do it, but not everyone can prove it. Therefore, if you grant that the most relevant and scientific part of such propositions is mine, it's not unfair to say that those propositions themselves are my own, especially since their form has no comparison with that of Giordano's. Each time someone produces a piece of work with a different order and structure from those of another author, even if its content is almost the same, he can reasonably claim it as his own,

because competence is more adequately judged by order and organization than by the level of the subject under consideration.[43]

After reading Tartaglia's plea, and based on his argumentation, one cannot fail to notice that Cardano could have claimed the paternity of the rules for solving cubic equations. On this point, let us hear historian Bortolotti:

> It seems to me that Cardano could not have found a better apologist. In fact, after having obtained from Tartaglia only the statements of the propositions, he produced his own demonstrations and completed, broadened, and organized the subject matter within a logically connected system. According to Tartaglia's reasoning, Cardano could have *claimed those propositions as his own*[44]; instead, he openly and repeatedly named their first discoverers, and only gave himself the credit that was his due.[45]

After having confirmed his rejection of an oral debate, Tartaglia attached a list of thirty-one problems, giving his rival—or rivals, since he was addressing himself also to Cardano—fifteen days to find the solutions. The first seventeen were problems about geometric constructions with ruler and compass, followed by three others on mathematical geography based on Ptolemy's book *Geographia*; the twenty-first was a problem on solid geometry; the remaining ones, up to the thirtieth, involved the extraction of arithmetic roots, and only the last one was about cubic equations.

Ferrari's third pamphlet was divided into two parts, dated respectively May 24 and June 1, 1547. In it, the Bolognian mathematician not only did not answer any of the questions posed by Tartaglia but dismissed them as "... useless, although not totally uninteresting as exercises; they should be not at the top, but at the bottom of a contest that everyone in Italy is waiting for.[46]

Soon afterwards Ferrari reiterated his proposal of a "real confrontation," that is, a face-to-face meeting, equating Tartaglia's eventual refusal of such format to his turning down the challenge. In the second part of the pamphlet, Cardano's disciple sent back thirty-one problems to

Niccolo, indicating that—just as those proposed by his rival—they should be considered only as an appendix to the oral duel he was hoping for. Significantly more difficult than Tartaglia's questions, the list was composed of seven problems in algebra, three of which led to solving third-degree equations; thirteen problems in geometry, seven in astronomy, three in philosophy, and one in architecture. In the words of historian Bortolotti:

> It is immediately clear to anyone examining the statements of the problems that, by the level and variety of the topics, ingenious formulation, and textual elegance, they are entirely different from those proposed by Tartaglia. Some of them reach the very limits of the algebraic theories of the time; others require geometric intuition and mastery of calculation techniques, and yet others humanistic erudition. None of them, however, is beyond the notions and possibilities that, by the discoveries made and communicated, were familiar to the practitioners of science, and thoroughly familiar to Ferrari.[47]

There is no doubt that Niccolo couldn't have encountered a more formidable opponent.

The exchange of pamphlets between Ferrari and Tartaglia continued from June to October 1547, both contenders impassively carrying on with their accusations and recriminations. In short: while Tartaglia persisted in denouncing Cardano's dishonor by having gone back on his oath, and seeking by every means to draw him into the controversy, Ferrari continued to maintain that in *Quesiti* his rival had distorted and misrepresented the facts and the documents concerning his diatribe against Cardano, and emphasizing the reality that because it was he, Ludovico, alone who had launched the challenge, the contest was only between the two of them.

They also discussed at length and in great detail the solutions to their respective problems. Tartaglia even proclaimed himself the winner for having answered Ferrari's questions in less time than it took the Bolognian to solve his. The pamphlets confirm, however, that Ferrari's solutions were complete and indisputable—even if some were not accepted by his rival, who had worked out more clever solution methods[48]—while not all of Tartaglia's solutions were really satisfactory. Moreover, Niccolo was still reluctant to conclude the challenge with an oral confrontation, as his adversary firmly requested.[49]

Ferrari's pamphlets and Tartaglia's counter-pamphlets followed one another in rapid succession up to the fifth, which Ludovico issued in October 1547. After that, nearly eight months went by without any reply from Tartaglia, who was thought to have abandoned the contest. Suddenly, Niccolo wrote again in June 1548, invoking a period of poor health and a return to his hometown, Brescia, three months earlier as the reasons for his silence. But, above all, after his long and stubborn resistance, Tartaglia surprisingly accepted the "real confrontation"—addressing the counter-pamphlet, his fifth, to the Ferrari–Cardano couple:

> Messer Hieronimo and Messer Ludovico, you remind me of Astolph of England, who claimed to be the best rider in the world but in fact always ended up flat on the ground and blamed his horse for it. You suffered the same fate during our mutual challenges, and presently I'm going to push you into a corner from which you will not easily escape. To get to the point, I want to tell both of you that I gladly accept a real confrontation with you. And since I happen to be not far away, I'll spare you the inconvenience of having to travel to Rome, or Florence, or Pisa, or Bologna, for I've decided to come in person to Milan.[50]

Why did Niccolo change his mind?

We can speculate on his motives by reading the sixth and last of Ferrari's pamphlets, dated July 14, 1548. Ludovico warned his rival that he knew the true reasons behind his change of heart, and he was probably right on target. In short: Tartaglia had been invited back to Brescia by some of the city's notables, who had offered him a chair of geometry with a considerable remuneration; it was likely, however, that a specific condition for awarding him the position was attached, namely, that Niccolo conclude as soon as possible the well known and lengthy duel with Ferrari—with him as the victor, of course.

For his part, Ferrari responded with his usual brio to the boldness shown by his rival in resuming the hostilities:

I don't know where you picked up the idea that I remind you of Astolph, who brags and then falls flat on his back. Because I only entered into contests with you, and on another occasion with your friend, you know who [Tonini de Coi], and each time what happened was the exact opposite of what you say. In the case of your friend, in front of the entire city and excellent judges, he was clearly defeated and went away, while I was chosen for the position in Milan. And so far I have treated you in such a way that you left Vinegia for Brescia, which, even if it's a most honorable city, I doubt that next time you will go from Brescia to a place worthy of someone as learned as you. Thus, let me ask you: Are these events an indication that I fell, or that I made someone else fall on his back?[51]

And so, in a malicious and unstoppable crescendo, the young Bolognian showered Tartaglia with epithets such as "beastly creature," "devil," and "snake head," adding that with his pamphlets he had "crushed his backbone" and urged him to summon the little strength he had left for the final confrontation, for otherwise he would have to spend the rest of his days "in a mire of ignominy, as the personification of ignorance and malevolence."[52]

Finally, he reminded him of Cardano's kind gesture—having cited him in *Ars magna*:

Even if this doesn't concern me, I think you condemn yourself, since Signor Hieronimo could have attributed that rule to its first discoverer, Scipione Dal Ferro, or to Antonio Maria Fior who, as you admitted in your book, knew it before you; even so, he was so kind as to believe you had found it on your own, without learning it from them or their students, and has praised you together with both of them. And you, instead of appreciating his benevolent gesture and others I mentioned to you in my second pamphlet, have inappropriately written so wickedly about His Signoria that you seem to have lost your mind.[53]

Tartaglia replied on July 24, 1548, with a counter-pamphlet addressed once again to both Ferrari and Cardano, in which he expressed his regret for the fact that the latter didn't want to participate in the oral contest. He nevertheless confirmed his intention to go to Milan to publicly discuss the solutions to his problems given by the young Bolognian.

Tartaglia's reply marks the end of the written phase of the great controversy. Ferrari's six pamphlets together with his rival's six responses constitute "one of the most extraordinary documents in the history of mathematics."[54] They provide a vivid, extremely valuable portrayal of mathematical knowledge in the middle of the sixteenth century by those at the top of the discipline—the two contenders and Cardano—and offer "a clear insight into the customs of the mathematical community of the time,"[55] which are also seen through the crude language of one of the most ferocious disputes in the history of science.

The moment of truth came on August 10, 1548, when Ferrari and Tartaglia crossed mathematical swords in Milan, at the Church of Santa Maria del Giardino, then run by Franciscans of the Order of Friars Minor.[56] A huge crowd was present, including all of the city's notables with the exception of Cardano, who had chosen to go away for the duration of the debate. Sitting in the quality of supreme judge was don Ferrante Gonzaga, who two years earlier had succeeded Alfonso d'Avalos as governor of Milan. The public duel would turn on the sixty-two problems contained in the pamphlets and their respective solutions.

Unfortunately, no formal records of the Milan dispute or of its final outcome exist. On the other hand, there are some after-the-fact accounts from Cardano and Tartaglia that, unsurprisingly, significantly disagree with each other. Gerolamo devoted a brief but peremptory note to the event, declaring that Tartaglia was defeated and forced to "sing the palinode," that is, retract the accusations against him.[57] More detailed and extensive, besides being different, is Niccolo's account, which appears in his previously mentioned *General trattato di numeri, et misure* (General Treatise on Number and Measure), published in 1556. He begins as follows:

> While I was in Brescia, not far from Milan, I decided (in order to put an end to the pamphlets, which by now only served to annoy people) to go to Milan and let them [i.e., Ferrari and Cardano] know in

public, loud and clear, that their solutions to my problems were for the most part incomplete or wrong. I therefore rode my horse to the city and put up some public posters[58] inviting them to come, on Friday, 10 August 1548, at hour 18, to the temple known as the Garden of the Zoccolanti Friars [that is, the begging friars], so that we could debate on my objections to their solutions, received seven months past the deadline set to answer the thirty-one questions I proposed.[59]

But Cardano, as we have already mentioned, did not show up. And while Ferrari came with a large number of supporters, Niccolo was accompanied only by his brother Zuampiero.

Hieronimo Cardano, in order to avoid the confrontation, suddenly left the city, so that only Ludovico Ferrari showed up. He was accompanied by a large following of supporters and friends, and I only by my brother, who had come with me from Brescia. In front of that crowd I briefly explained the nature of our dispute and the reasons for my presence in Milan.[60]

After enduring a quarrel over the judges of the duel that lasted two hours, Tartaglia could finally begin to raise his objections to Ferrari's—or rather, according to him, to Ferrari's and Cardano's—solution to his eighteenth problem, drawn from Ptolemy's *Geography*, but the compact crowd soon started to manifest signs of hostility, preventing him from proceeding with his argumentation:

I was ready to expose their false solutions to my problems but the crowd, to disrupt my purpose, wasted more than two hours trying to appoint as judges certain persons present who were friends of his [Ferrari's] but unknown to me. I did not agree to such an astute subterfuge, and told them I wished all those present to be the judges and, similarly, all those who would receive my arguments in print, and so they finally let me speak. Since I wanted to avoid annoying many noble listeners with difficult topics in arithmetic or geometry, I decided to begin with my objections to the answer they gave to my eighteenth problem, based on the twenty-fourth chapter of Ptolemy's *Geographia*, and thus force them to admit that their solution was wrong. But the audience loudly demanded that I should also let him speak about my solutions to the 31 problems he had sent me (which took me about three days to solve). I shouted back, and pleaded with them to first let me finish what I had to say, but to no avail; despite my complaints, they wouldn't hear another word from me and insisted I let Ferrari speak.[61]

Emboldened by the support of his "large following," Ferrari took the floor to attack his rival's answer to his fourth problem, a question in architecture about the Latin architect and engineer Vitruvius' rules to determine the relationship between the dimensions of the various parts of a building. It was a minor subject, considering the ambitious list of topics that appeared in the pamphlets. But he could never finish his argument, because the contest rapidly came to a confusing and disappointing end:

> Ferrari began by claiming that I hadn't solved his fourth problem on Vitruvius, but his argument was so long that at dinner time the crowd vacated the temple and everybody went home. I realized that in such a place I wouldn't be able to make my case because Ferrari had the crowd completely on his side, and I began to fear the worst. And so the next day I went discreetly back to Brescia along a different path from the one I took when I came over to Milan, having decided to make public in a printed document what they wouldn't let me say in person.[62]

Thus, after only one day of debating, and long before the most serious questions could be discussed, Niccolo abandoned the contest, which in his view had been invalidated from the start by the public's behavior, as everyone—or almost—had noisily taken Ferrari's side. In the absence of official records and given the discrepancy between Cardano's and Tartaglia's accounts, we are left in the dark as to how the duel really unfolded. Perhaps there was no "objective conclusion,"[63] and maybe Niccolo was actually the victim of Ferrari's supporters' hostility. In the meantime, it is legitimate to conjecture that Tartaglia's sudden return from Milan was nothing else than a way to cut his losses from what he perceived as an almost certain defeat, not on account of the spectators but rather because of the ability and preparation of a rival he had probably underestimated. Various historians consider Ludovico Ferrari "the sharpest and most profound mathematician of his time,"[64] even superior to Tartaglia and Cardano. Certain facts in the biographies of the two contenders seem to indicate that for his contemporaries Ferrari was the victor of the duel. In particular, not long after the contest Ludovico received numerous and enticing work offers, among which was a public teaching position in Rome, a private one in Venice, and, perhaps, even an invitation to serve the Emperor Charles V of Habsburg as preceptor of his son. He finally accepted, against Cardano's advice, the position of director of the Duchy of Milan's land registry, offered to him by the governor, Ferrante Gonzaga. Niccolo, for his part, was dismissed from his teaching post in Brescia, without compensation for the classes he had already taught, and went back to his humble occupation as abbaco master in Venice, deeply worried and without a cent. In his life, he had fought and won many mathematical duels, but the last one destroyed him.

Different, but equally steeped in bitterness, were the destinies of the protagonists of this story.

Niccolo Tartaglia died in Venice on December 13, 1557, in solitude and poverty. The previous year he had published the first two parts of *General trattato di numeri, et misure,* his major work, whose four remaining parts would be released posthumously in 1560. Following the path marked out by Leonardo Fibonacci's *Liber abaci* and Luca Pacioli's *Summa, General tratatto* is a monumental and lively mathematical encyclopaedia, widely circulated after its author's death and greatly appreciated. In its pages appears the famous array of numbers known in Italy as "Tartaglia's triangle,"[65] familiar to every student of mathematics.[66]

In *General trattato,* besides telling his own version of the duel against Ferrari, Niccolo gave the solutions to the problems proposed by the latter during their exchange of pamphlets and counter-pamphlets, pointing out the "errors" and "blunders"[67] made by his rival (or rivals, given that he kept mentioning also Cardano). In fact, they were not strictly speaking errors or blunders but solution methods different from his. The chapter devoted to algebra does not include the theory of cubic equations. Niccolo had intended to discuss it in a later part of the work, but he didn't have the time.

Ludovico Ferrari quit his position at the Milan land registry in 1557 for health reasons, and went back to his native Bologna to live with his sister Maddalena, "widow and impecunious."[68] There he resumed his studies, humanistic as well as scientific, and in 1564 was appointed to the mathematics chair at the local university. Death came suddenly in October of the following year. He was only forty-three when he passed away, probably poisoned by his sister.

> Even if there is no proof of the presumed poisoning, Maddalena's behaviour confirmed her sinister design: she married two weeks after her brother's death and transferred all the possessions Ferrari had bequeathed to her in his testament to her husband. It appears that Cardano went to Bologna—during Ferrari's last days or shortly after his death—to recover whatever was left of his books and writings. Unfortunately he arrived too late, since Ferrari's brother-in-law had already secured possession of his latest works on Cesar and Vitruvius with the intention of having them published under the name of a son from a previous marriage, and reap the profits.[69]

In fact, the solution of the fourth-degree equation—which Cardano included in his *Ars magna*—and the answers to Tartaglia's questioms in the pamphlets of the mathematical duel were the only works that Ferrari had the satisfaction of seeing published.[70]

Thanks to his numerous books on every field of the knowledge of his time and his skill in the practice of medicine, Gerolamo Cardano achieved fame and recognition in the mature years of his life. In the late 1540s, for instance, both Pope Paul III and King Christian III of Denmark offered him alluring positions as physician of the court, which he declined. In 1552 he accepted instead to travel to Edinburgh, where he successfully cured Scotland's archbishop John Hamilton of asthma, receiving a considerable sum of money for his services. Back in Milan in January of the following year, he declined invitations from the king of France Henry II and Scotland's Queen Maria de Lorena.

In the midst of a successful and gratifying professional career, Cardano's life was abruptly devastated in 1560 when his eldest son Giovanni Battista was executed for poisoning his unfaithful wife. Overcome with grief and preoccupied with certain rumors being circulated about him, in 1562 Gerolamo left Lombardy for Bologna, where he held the chair of medicine at the university. A few years later he was constrained to report to the police his second and dissolute son, Aldo, who had stolen money and jewelry from him after repeated losses playing dice. Like father, like son, but only in vice. As punishment for his deed, Aldo was expelled from the city.

Between 1570 and 1571, the father himself went through a troubling period; Gerolamo spent several months in prison and house arrest serving a sentence for heresy handed down by the Inquisition. Although the charges against him by the Holy Office are not known, they might be related to his frequenting Protestant circles,[71] or to some impertinent passages of his writings, such as his Horoscope of Jesus Christ or Nero's eulogy. In September 1571, after recovering his freedom at the price of his recantation and commitment not to hold public lectures or publish other books, he moved to Rome, and two years later was granted a pension by Pope Gregory XIII. In Italy's capital, Cardano led a retired life, occupied mostly with writing his memoirs. He died on September 20, 1576, almost three years later than the date—December 5, 1573—he had predicted for the inescapable event.

In his autobiography, published in Paris in 1643, Gerolamo acknowledged once again Tartaglia's mathematical contributions, although in a bittersweet manner, and unambiguously asserting his own merit:

> I began working on the *Ars magna* at the time of the disputes with Zuanne da Coi and Tartaglia, who had sent me the first equation. But he preferred to acquire an enemy—one superior to him—rather than conquer a friend who has a debt to repay, even if the formula wasn't his.
>
> I admit that I owe brother Niccolo some mathematical discoveries, but not many, and they don't diminish those I made myself.[72]

In the first half of the sixteenth century, Scipione Dal Ferro, Niccolo Tartaglia, Gerolamo Cardano, and Ludovico Ferrari were the four glittering musketeers who lit up the sky of algebra with their extraordinary and far-reaching discoveries. These were the result not only of their creative genius and technical skill, but also of passion, dedication, perseverance, rivalry, jealousy, ambition, esteem, resentment, impetuosity, and suffering—in a word, of all the human emotions and feelings that can be hidden in a mathematical formula.

NOTES

CHAPTER 1 The Abbaco Master

1. Known today as the "Falcon of Italy," the Castle of Brescia is located on the 250-meter-high Cidneo Hill that dominates the city's historic center. For further information on this impressive and beautiful fortress, one of the best preserved in northern Italy, see for example [3].

2. [2], p. 270.

3. In the absence of indisputable archival sources or explicit information from Tartaglia himself, the year of his birth cannot be established with certainty. The hypothesis that Niccolo Tartaglia was born in 1499 is based on two 1529 civil register booklets of the Veronese district of Santa Maria Antica (as we shall see, Niccolo was living in Verona at the time). These documents also report Tartaglia's family status; his age, thirty years old, is inscribed next to his name. The booklets, bearing the numbers 586 and 587, are presently in the Verona State Archives. Even if in those days registers were often inaccurate, the year of Tartaglia's birth deduced from the Veronese booklets is roughly in accord with the estimate gleaned from his memoirs (around 1500).

4. [100c]. Tartaglia gives here an account of the brutal assault he suffered by a French soldier during the 1512 sack of Brescia (*Libro* VI, *Quesito* VIII, f. 69v).

5. A native of Martinengo, in the province of Bergamo, Gabriele Tadino was a skillful military engineer and man of arms who lived between ca. 1480 and 1543. His heroic participation in 1522 in the defense of the Greek island of Rhodes besieged by the Turks earned him glory and honors; he fought under the banner of the Knights of Malta and was seriously injured in one eye. A portrait of him, from the collection of the Ferrara Savings Bank, is attributed to the famous Venetian painter Tiziano Vecellio. For information on Tadino's life and in particular his relationship with Tartaglia, see [97] and [71e] respectively.

6. [100c], f. 69r.

7. [100c], f. 69r. On the basis of the available information, it is not possible to explain why in this extract Tartaglia mentions only one sister, while in the previous one he refers to two. In his testament (see next note) Niccolo mentions a sister named "Catharina."

8. In other passages of the same document, "Zuampiero" becomes "Zampiero." Tartaglia's testament, drawn up by the notary Rocco De Benedetti on December 10, 1557, and now in the Venice State Archives, can be found in [71c], Tables XXXI–XXXIII.

9. See [44b], pp. 13–15.

10. [100c], f. 69v.

11. This hypothesis is put forward by Luigi Bittanti in [18], p. 5. Tartaglia never revealed his mother's name, or any other information about her.

12. [100c], f. 69v.

13. Idem. On the "mercantesca" script see, for example, [29].

14. [100c], ff. 69v–70r.

15. For a detailed description and analysis of the characteristics and organization of Italian schools during the Renaissance see [56].

16. [100d], parte II, f. 27v.

17. [100c], f. 64r. The Italian word in the text, *gargione* (translated as "bachelor"), may refer to a person who performs particularly simple tasks, as well as to an unmarried man.

18. [100d], parte III, f. 7r.

19. [103b], p. 403.

20. Idem.

21. [56], p. 307.

22. Idem.

23. [211], p. 5.

24. [45a], p. 174.

25. [103b], p. 410.

26. [56], p. 309.

27. Known also as Leonardo Pisano.

28. As there are no copies of the first version, our knowledge of the text is based on the second version, from 1228. The first modern edition was published by Baldassarre Boncompagni in 1857 on the basis of a single medieval codex from the fourteenth century (known as *Maglia-bechiano*). It consists of the Latin text with no commentaries. No Italian version of this important work exists, but it has recently been translated into English from the Boncompagni Latin edition. See [37].

29. [90].

30. Our usual number system is called positional because the value of each digit depends on the position it occupies in the sequence that make up the number. For example, in 124, the digit 2 indicates two times ten, while in 278 it indicates two times one hundred. The Roman system, on the other hand, is additive; that is, symbols always keep the same value regardless of the position they occupy (X always represents 10; L always stands for 50).

31. Here and in what follows, the term "algebra" will be understood in its classical and elementary sense, that is, as the study of the calculus with letters and equations using the arithmetic operations.

32. For precision's sake, it should be noted that abbaco treatises referred also to another of Fibonacci's works, *Practica geometriae* (The Practice of Geometry), published by the Tuscan mathematician in 1220. Further models for these treatises were perhaps a lost work by the same author, *Liber minoris guise* (Book in a Smaller Manner), which presumably constituted a reduced and simplified version of the *Liber abaci* (see [40e]), as well as possible sources of Arabic tradition coming from the Ibero-Provençal environment and different from those used by Fibonacci (see [Jens Høyrup, "Leonardo Fibonacci and *abbaco* culture.

A proposal to invert the roles," *Revue d'histoire des mathématiques*, vol. II, 2005],
pp. 23–56).

33. [103b], p. 415.

34. Idem.

35. For a summary of Piero Della Francesca's mathematical interests and works see [45b].

36. See, for example, [7].

37. [56], p. 316.

38. [45a], p. 178.

39. Idem.

40. [100d], parte I, f. 148r.

41. In the *Liber abaci*, for example, the solution to a problem in recreational mathematics—
involving the reproduction of rabbits—leads the author to the famous sequence of integers
known as the "Fibonacci sequence": 1, 1, 3, 5, 8, 13, 21, 34, 55, 89, . . . (each number, except the first
two, is the sum of the two numbers immediately preceding it). Over time, a great number of
surprising applications of the Fibonacci sequence were found in a variety of fields: mathematics,
natural sciences, economics, biology, and sociology, among others.

42. [48], p. 315.

43. These are the same documents as those mentioned in note 3.

44. See [44b], pp. 24–25. We remind the reader that in the monetary units then in use, one
lira was equivalent to 20 soldi.

45. The incident is reported in *Atti dei rettori veneti*, kept in the Archives of the State of Verona
(years 1532–33, n. 43: "Parte Testium Tertium," ff. 67–69). See also [34], p. 353.

46. [100d], parte I, f. 171v.

47. [100c], f. 4r.

48. [21e], p. 165.

49. Idem.

50. At that time, "lecturer" was equivalent to "professor."

51. [21e], p. 163.

52. Idem.

53. *Coi* is the dialect form of the toponym Collio, a small town in Val Trompia, Province of
Brescia, which is clearly the mathematician's place of origin. *Zuanne* is also a dialect version of
the name Giovanni.

54. [100c], f. 101r. In both problems, the numbers to be found are assumed to be positive.

55. [82].

56. Idem, f. 149r.

57. Idem, f. 150r.

58. Equations. More precisely, the mathematicians of the epoch considered the various types
of equations as the "chapters" of algebra.

59. [100c], f. 101r.

60. Idem, f. 101v.

61. Idem.

62. Idem.

CHAPTER 2 The Rule of the Thing

1. [70e], p. 11.

2. The archives also contained tablets inscribed in Sumerian, and the most ancient bilingual (Sumerian-Eblatic) lexicons known.

3. [70e], p. 13.

4. The Sumerian city of Kish held close cultural ties with Ebla, to the extent that the cuneiform alphabet used in Ebla conforms to stylistic rules proper to the one employed in Kish.

5. Livia Giacardi, "Sistema di numerazione e calcolo algebrico nella terra tra i due fiumi," in [4], p. 64.

6. [105].

7. [84a], p. 192.

8. Quoted in Clara Silvia Roero, "Numerazione e aritmetica nella matematica egizia," in [4], p. 121.

9. The *hekat* was a unit of volume used in ancient Egypt (especially to measure volumes of grain), equivalent to approximately 4.8 liters.

10. Our paraphrase of the problem's original text—in fact, a table of data (in which the answer to the question is also given). The problem and an analysis of it appear in "Numerazione e aritmetica nella matematica egizia," in [4], pp. 137–39.

11. Quoted in [53]. In Fibonacci's version, there is one more power of seven (7^6) than in the problem of the Rhind papyrus. The answer is 137,256.

12. Quoted in Clara Silvia Roero, "Numerazione e aritmetica nella matematica egizia," in [4], p. 137. The solution is $7 + 7^2 + 7^3 + 7^4 = 2{,}800$.

13. See [70e], p. 551.

14. [41a], p. 25.

15. Livia Giacardi, "Sistema di numerazione e calcolo algebrico nella terra tra i due fiumi," in [4], pp. 45–46.

16. Idem, p. 46.

17. Idem.

18. [78], p. 67.

19. [23], p. 30.

20. Livia Giacardi, "Sistema di numerazione e calcolo algebrico nella terra tra i due fiumi," in [4], p. 66.

21. Quoted in [101], p. 21. Numbers are written in our system—not, as in the quoted source, in that of the Babylonians.

22. It will also appear toward the end of this chapter.

23. [95], p. 15.

24. It is easily checked that the equation also admits the solution $x = \text{-}7$, which, being negative, has no geometrical meaning.

25. On the other hand, there is no doubt that the Greeks were familiar with the results obtained by the Babylonian mathematicians. In this respect, see [70e], p. 110.

26. Idem, p. 111.

27. [41a], p. 21.

28. Our translation of the Latin version of *Anth. Pal.* XIV, 126: "*Egit sextantem juvenie;/ lanu-gine malas vestire hinc coepit parte duodecima / Septante uxori post haec sociatur,/ et anno formosus quinto nascitur inde puer./ Semissem aetatis postquam attigit ille paternae,/ infelix subita morte peremptus obit./ Quator aestater genitor lugere superstes cogitur,/ hinc annos illius assequere.*" Fred Dubner, Edme Cougny, Maximus Planudes, Hugo Grotius, *Epigrammatum anthologia Palatina*, Ambrosio Firmin-Didot, Paris 1864–1872.

29. Diophantus addressed in a systematic way the study of equations with multiple un-knowns that may have an unlimited number of solutions (indeterminate equations), thus creat-ing a branch of mathematics later called Diophantine analysis. We shall not discuss here the results he obtained on indeterminate equations nor their development, given that Diophantine analysis is usually not considered as part of algebra but rather of number theory.

30. [63], p. 139.

31. Idem.

32. "Viète introduced what may be considered the most sweeping innovation in algebraic symbolism, that is, the systematic use of letters to indicate both unknowns and constant terms," [41a], pp. 10–11.

33. Idem, p. 9.

34. [70e], p. 130. In the case of quadratic equations with two positive solutions, Diophantus considered only the largest one, and refused to accept negative or irrational solutions.

35. [90].

36. [62a].

37. Quoted in [94], p. 97.

38. [23], p. 227.

39. For a modern version of this important work, with Arab text and English translation, see [62b].

40. [90].

41. [23], p. 230.

42. [90].

43. "Algebra and almucabala, commonly known as Rule of the Thing," [100c], f. 1v.

44. The constant term of an equation is the number that occurs "by itself"; for instance, in the equation $x^3 + 4x = 21$, the constant term is 21.

45. See [70e], p. 552, and [41a], p. 8.

46. [90].

47. The case of the first-degree equation $ax = c$ is considered separately, as its solution re-quires only a simple division ($x = c/a$).

48. [90].

49. [70e], p. 547.

50. [63], p. 220.

51. [32].

52. [90].

53. See, for example, [61], and the more recent [89].

54. As before, Khayyam's verbal expressions are listed on the left—his algebra was also purely rhetorical—and the corresponding modern notation on the right.

55. The equation was proposed to him by the scholar Johannes of Palermo during one of the mathematical duels that opposed them in Pisa, as we mentioned in the previous chapter.

56. Quoted in [90].

CHAPTER 3 The Venetian Challenge

1. Quoted in [34], p. 356.

2. On this topic, see, for example, [9], pp. 193–96.

3. This information comes from Carlo Maccagni, [67], p. 485.

4. Idem, p. 489.

5. In classical antiquity, the seven "liberal arts"—grammar, rhetoric, dialectics, arithmetic, geometry, astronomy, and music—were so called because they were practiced by free men, that is, those devoted exclusively to intellectual activities, as opposed to "non-free" men, who performed mostly manual labors ("mechanical arts").

6. Quoted in [44b], p. 30.

7. *Apollonii Pergei philosophi mathematicique excellentissimi Opera. Per doctissimum philosophum Ioannem Baptistam Memum Patritium Venetum, mathematicharumque artium in Urbe Veneta lectorem publicum de Graeco in Latinum traducta et noviter impressa*, per Bernardinum Bindonum, Venetiis 1537.

8. [100c], ff. 104v and 118v.

9. Idem, f. 114v.

10. [91], p. 145.

11. [69], f. 8v.

12. [11b], p. 92.

13. Idem, p. 93.

14. [92].

15. The date of February 22, 1535, is explicitly mentioned in [100c], f. 114r, where there is a copy of the notarial document containing Fior's problems. The inscribed year, however, is 1534; this is not a clerical error but a consequence of the fact that in the calendar of the Republic of Venice, the New Year began on March 1.

16. [100c], ff. 106r—106v.

17. Idem, f. 106v.

18. We remind the reader that in the mathematical terminology of the time, *chapter* meant "equation," or "type of equation."

19. Taking into account the precise date Tartaglia mentions in the passage immediately following, and knowing that the problems were handed to the notary on February 22, 1535, those "eight days" should actually have been "ten days."

20. See note 15.

21. [100c], ff. 106v.

22. Idem.

23. [100c], ff. 106v.—107r.

24. Idem, f. 107r.

25. In the original Italian, the man was of Jewish origin. In those times in Europe, money lending was one of the few activities Jewish people were allowed to exercise, and one that was forbidden to Christians.

26. [100c], ff. 114r—114v. The numbering of the problems is the same as the one used by Fior.

27. The equation arising from the second problem can be reduced to the same form by dividing each member by 4.

28. [100c], f. 107r.

29. Idem.

30. Idem.

31. Idem.

32. [100c], f. 107v.

33. Idem.

34. Idem.

35. Idem.

36. Idem.

37. Idem, f. 108r.

38. Idem.

39. Idem, f. 103v.

40. The problem that led to the fourth-degree equation.

41. That is, the four equations listed in the text and resulting from problems Tartaglia addressed to Fior.

42. [100c], f. 108r.

43. Tartaglia is still referring to the problem Tonini da Coi had anonymously sent to him to be solved with a fourth-degree equation.

44. [100c], ff. 108r–108v.

45. Idem, f. 108v.

46. Idem.

47. Readers interested in technical details not discussed here may find them in [70c].

48. On this topic, see [21i], pp. 47–48; [70c], pp. 279–83; Arnaldo Masotti, "Niccolo Tartaglia e i suoi 'Quesiti'," in [71c], p. 26; [77], p. 637.

49. [70e], p. 248.

50. We report here a passage from [21e], p. 171.

51. [100c], f. 110r.

52. Idem, f. 111v.

53. Idem, f. 110v.

54. Idem, f. 111r.

55. In another passage of his January 8, 1537, letter, Tonini da Coi wrote, in regard to Fior: "[. . .] because you [Tartaglia] sent me the solution of that problem of yours, I did not want to go back and ask him [Fior] for it, so as not to be indebted to him," idem.

56. Idem.

57. Idem.

58. Idem, f. 110v.

59. Idem.

60. Idem, f. 112r.
61. Idem, ff. 112r—112v.
62. Idem, f. 112r.
63. Idem, f. 112v.
64. [100a].
65. [58], p. 587.
66. [100b], f. 2r [n.n.].
67. Idem.
68. Idem.
69. [68], p. 239.
70. [100c], f. 113r.

CHAPTER 4 An Invitation to Milan

1. [26e], p. 37.
2. Idem, p. 39.
3. Idem.
4. Idem, p. 40.
5. Idem, p. 41.
6. Idem, p. 115.
7. [54], p. 759.
8. [26e], p. 60.
9. Following its defeat by the French in the battle of Pavia in February 1525, Milan was governed by Francesco II Sforza under the protection of Charles V of Habsburg, King of Spain and Holy Roman Emperor. In 1535, after Sforza's death, Charles V took power, beginning Spain's direct domination over the city and the duchy that would last 170 years.
10. [26e], p. 76.
11. A byproduct of Cardano's passion for gambling is the small treatise *Liber de ludo alea* (The Book on Games of Chance) posthumously published in 1663. It contains the first study on the foundations of probability. For a recent edition of this important work see *The Book on Games of Chance: The 16th Century Treatise on Probability*, by Samuel S. Wilks (Foreword), Gerolamo Cardano, Sydney Henry Gould (Translator), Dover Publications, 2015.
12. [26e], pp. 64–65.
13. Idem, p. 101.
14. Idem.
15. Idem, p. 92.
16. Idem.
17. Idem.
18. Idem, p. 84.
19. [54], p. 759.
20. [26e], p. 84.
21. Idem, p. 120.
22. Idem, p. 126.

23. Idem, p. 127.

24. [81], p. 33.

25. Idem, p. 34.

26. [26a].

27. [27], p. 13.

28. [26e], p. 43.

29. [76], vol. 1, pp. 148–49.

30. [100c], f. 113r.

31. Idem.

32. Idem, ff. 113r–113v.

33. Idem, f. 113v.

34. The equation in question is $x^4 + 8x^2 + 64 = 160x$ (see p. 52).

35. The equation in question is $x^3 + 40x^2 = 2888$ (see p. 54).

36. [100c], f. 113v.

37. Alfonso d'Avalos, governor of Milan.

38. [100c], f. 113v.

39. Idem, ff. 113v–114r.

40. [100c], f. 115r.

41. Tartaglia's *Nova scientia*.

42. Alfonso d'Avalos, Marquis del Vasto and governor of Milan.

43. [100c], f. 115r.

44. Idem.

45. [100c], ff. 115r–115v.

46. Idem, f. 115v.

47. Idem.

48. Idem.

49. Idem, f. 116r.

50. Idem.

51. Idem.

52. Idem, ff. 116r–116v.

53. Idem, f. 116v.

54. Idem.

55. [81], p. 71.

56. Idem.

57. The Venetian typographer Ottaviano Scotto was Cardano's first publisher. He had already published in 1536 the previously mentioned *De malo recentiorum medicorum usu medendi libellus*. The two men developed and maintained a close and friendly relationship.

58. [100c], ff. 116v–117r.

59. Idem, ff. 115v–116r.

60. Moreover, the cubic equation in question is an example of the *irreducible case* that we shall discuss later on.

61. "Trappola" (Trap) was a card game played at the time. See [81], p. 71.

62. A trick played in the Middle Ages and the Renaissance. See idem.

63. [100c], f. 118r.

64. Idem.

65. Also "San Zanipolo," the name in dialect of the Venetian basilica of Saints John and Paul.

66. [100c], ff. 118r–118v.

67. Idem, f. 119r.

68. Idem, ff. 119r–119v.

69. That is: I rather live as a coward than die a hero.

70. [100c], f. 119v.

71. Idem.

72. Idem.

73. Idem, f. 120r.

74. Idem.

75. Idem.

76. Idem, ff. 120r–120v.

77. Idem, f. 120v.

78. Idem.

79. Idem, ff. 120v–121r.

CHAPTER 5 The Old Professor's Notebook

1. Saturday, April 5, 1539.

2. [100c], f. 121r.

3. See footnote on p. 89.

4. [100c], f. 121r.

5. Idem.

6. Idem.

7. Idem.

8. Idem.

9. Idem, ff. 121r–121v.

10. See footnote on page 89.

11. We remind the reader that we refer only to real solutions, since—as we shall see later—complex numbers were introduced only after the middle of the sixteenth century.

12. Idem, f. 121v.

13. That is, the value of x.

14. $u = v + c$ follows from the first equation of the system.

15. [21i], p. 49.

16. See Chapter 3, note 15.

17. This hypothesis is proposed in [70a], pp. 24–31.

18. [100c], ff. 121v–122r.

19. Idem, f. 122r.

20. Idem.

21. Idem.

22. [26b].

23. Quoted in [74], p. 68.

24. Idem.

25. [26e], p. 45.

26. [100c], f. 122r.

27. Idem.

28. Idem.

29. Idem, f. 122v.

30. Idem.

31. Idem, ff. 122v.–123r.

32. Idem, f. 123r.

33. Idem.

34. Idem.

35. Idem, f. 123v.

36. Idem.

37. Idem, f. 124r.

38. Idem, ff. 124r–124v.

39. [21i], pp. 51–52.

40. [100c], f. 124v.

41. Idem.

42. [26e], p. 145.

43. [11a], p. 630.

44. We shall not elaborate on the method used by Ferrari to solve fourth-degree equations. The interested reader is referred, for instance, to [23], or [70e], pp. 302–11.

45. This document is presently at the Bologna University Library.

46. Quoted in [21i], p. 43.

47. Idem.

48. See, for example, [21b], p. 161–62, or [70e], pp. 236–38.

49. [22b], p. 66.

50. Quoted in [21i], p. 45.

51. Idem, pp. 45–46.

CHAPTER 6 The Final Duel

1. [70a], p. 174.

2. The book's original edition is: Gerolamo Cardano, *Artis magnae, sive de regulis algebraicis*, by Johannem Petreium excusum, Norimbergae 1545. This fundamental work has not yet been translated into Italian, but there is an English version, the most recent edition of which is: Gerolamo Cardano, *The Rules of Algebra (Ars magna)*, translated by Richard Witmer, introduction by Oysten Ore, Dover Publications, 2007.

3. Tartaglia published this book—the first printed edition of Euclid's *Elements* not only into vernacular Italian but also into any living language—in 1543. See: *Euclide Megarense philosopho, solo introduttore delle scientie mathematice: diligentemente rassettato, et alla integrita ridotto per il degno professore di tal scientie Niccolo Tartalea, brisciano,* Venturio Rossinelli ad instantia e

requisitione de Guilielmo de Monferra, et de Pietro di Facolo da Vinegia libraro, Venezia 1543. We remind the reader that in those times Euclid of Alexandria—*Element's* real author—was confused with the homonymous philosopher Euclid of Megara.

4. Muhammad ibn Musa al-Khwarizmi, considered by Cardano as the founder of algebra.

5. Italian translation taken from [70e], pp. 245–46.

6. Idem, p. 246.

7. "Real" solutions, in today's mathematical terminology.

8. [70a], p. 175.

9. See [20].

10. [70e], p. 450.

11. [22b], p. 71.

12. [100c], f. 125r.

13. Idem, f. 125v.

14. Idem.

15. Idem.

16. Idem.

17. Arlotto Mainardi, better known as "the piovano Arlotto," was a fifteenth-century Florentine priest famous for his practical jokes and witty remarks.

18. *De veris narrationibus* (The True Story), referred to by Ferrari in a satirical way, is an adventure novel by the second-century rhetorician Lucian full of fantastic elements and extravagant situations.

19. *Homeromastix* (Homer's whipper), was the nickname given to the Greek grammarian Zoilus (fourth century AD), an obstinate detractor of Homer.

20. [36], pp. 5–6.

21. Idem, p. 6.

22. On these ethical faults of the nevertheless great Brescian mathematician, see, for example, Arnaldo Masotti, "Niccolo Tartaglia," in [2], pp. 613–14.

23. A native of Granada, Spain, Diego de Mendoza, who lived between 1503 and 1575, was a politician and soldier with a strong interest in science and literature. From 1539 to 1546, he was emperor Charles V of Habsburg's ambassador to Venice, where Cardano introduced Tartaglia to him. Mendoza appears in Tartaglia's *Quesiti* in a discussion on statics.

24. [36], pp. 5–6.

25. Idem, pp. 6–8.

26. Idem, p. 17.

27. Idem, p. 18.

28. Idem, pp. 18–19.

29. [21h], p. 21.

30. [36], pp. 19–20.

31. Idem, pp. 20–21.

32. Idem, p. 21.

33. Arnaldo Masotti, Introduction, idem, p. XII.

34. Idem, p. 22.

35. [21h], p. 22.

36. Italian translation taken (with slight modifications) from [21d], p. 89.

37. A reference to the already mentioned Diego de Mendoza, emperor Charles V's ambassador to Venice.

38. [21d], pp. 89–90.

39. Idem, p. 90.

40. As we have already remarked, in spite of his claim, Tartaglia had not yet found any general rule for the solution of that type of cubic equation.

41. [36], pp. 43–44.

42. Idem, p. 44.

43. Idem, p. 45.

44. In italics in the text.

45. [21d], p. 93.

46. [36], p. 62.

47. [21h], pp. 34–35. An analysis—even brief—of the problems Tartaglia and Ferrari sent to each other is beyond the scope of this book. Readers interested in studying the subject in depth are referred to: Arnaldo Masotti, Introduction, in [36], pp. xxi–xxxiii; [83b] and [83c].

48. See [21h], pp. 75–76.

49. The historian Martin Nordgaard conjectured that Tartaglia's reluctance to accept a face-to-face confrontation was due to his difficulty in speaking ([79], p. 343). Although this possibility cannot be excluded, it is not known whether Niccolo's stammer, a consequence of the injuries he suffered in his youth, lasted all his life or was only temporary.

50. [36], p. 180.

51. Idem, pp. 187–88.

52. Idem, p. 188.

53. Idem.

54. [22b], p. 72.

55. Idem.

56. The church was located near Corsia del Giardino, presently via Manzoni, not far from the old La Scala opera house. It was knocked down in 1866.

57. See, for example, [26e], p. 182.

58. These documents are also lost.

59. [100d], f. 41v.

60. Idem.

61. Idem.

62. Idem.

63. [11a], p. 632.

64. [21i], p. 80.

65. [100d], ff. 69v and 71v.

66. In fact, "Tartaglia's triangle" was already known to Indian, Persian, and Chinese mathematicians at the end of the tenth century. In the middle of the seventeenth century, its properties were further studied by the French scholar Blaise Pascal, and almost everywhere it is now known as "Pascal's triangle."

67. [100d], f. 42r.

68. [11a], p. 632.

69. Idem, p. 633.

70. Other works by Ferrari, further evidence of the talent and versatility of the Bolognian mathematician, have been recently published in [39].

71. For example, Cardano had dedicated his *Ars magna* to the Lutheran theologian Andreas Osiander, known especially as the author of the anonymous foreword to *De revolutionibus orbium coelestium* (On the Revolutions of Heavenly Spheres), the famous 1543 treatise in which Nicolaus Copernicus introduced his heliocentric model of the cosmos.

72. [26e], pp. 164 and 203.

BIBLIOGRAPHY

1. AABOE Asger Hartvig. *Episodes from the Early History of Mathematics*. Washington, DC: The Mathematical Association of America, 1998.

2. AA.Vv. *Storia di Brescia*, edited by Giovanni Treccani degli Alfieri, vol. ii: *La dominazione veneta (1426–1575)*. Brescia: Morcelliana, 1963.

3. AA.Vv. *Il Castello di Brescia*. Brescia: Grafo, 1986.

4. AA.Vv. *L'alba dei numeri*. Bari: Dedalo, 1987.

5a. Acampora, Renato. *Die Cartelli di matematica disfida: Der Streit zwischen Nicolò Tartaglia und Ludovico Ferrari*. Munich: Institut für Geschichte der Naturwissenschaften, 2000.

5b. ———. "Nicolò Tartaglia und Ludovico Ferrari: die Cartelli." In Hartmut Roloff and Manfred Weidauer (eds.), *Wege zu Adam Ries. Tagung zur Geschichte der Mathematik (Erfurt 2002)*. Augsburg: Rauner, 2004, pp. 1–9.

6. Armocida, Giuseppe, Canziani, Guido, and Zanzi, Luigi (eds.), *Gerolamo Cardano nel suo tempo. Atti del Convegno, 16–17 novembre 2001 (Castello Visconti di San Vito, Somma Lombardo, Varese)*. Pavia: Edizioni Cardano, 2003.

7. Bagni, Giorgio T. *"L'arte de labbacho (L'Aritmetica di Treviso, 1478) e la matematica medievale."* In AA.Vv., *I seminari dell'Umanesimo latino 2001–2002*. Treviso: Antilia, 2002, pp. 9–32.

8a. Baldini, Ugo. "Cardano negli archivi dell'Inquisizione e dell'Indice: Note su una ricerca." *Rivista di storia della filosofia*, n. 4, 1998, pp. 761–66.

8b. ———. "L'edizione dei documenti relativi a Cardano negli archivi del Sant'Ufficio e dell'Indice: risultati e problemi." In Marialuisa Baldi and Guido Canziani (eds.), *Cardano e la tradizione dei saperi: Atti del Convegno internazionale di studi (Milano, 23–25 maggio 2002)*. Milan: Franco Angeli, 2003, pp. 457–515.

9. Barbier, Frédéric. *Storia del libro: Dall'antichità al xx secolo*, translated by Rita Tomadin, afterword by Mario Infelise. Bari: Dedalo, 2004.

10. Bellini, Angelo. *Gerolamo Cardano e il suo tempo*. Milan: Hoepli, 1947.

11a. Belloni Speciale, Gabriella. "Ferrari, Ludovico," *Dizionario biografico degli italiani*, vol. xlvi. Rome: Istituto della Enciclopedia Italiana fondata da Giovanni Treccani, 1996, pp. 630–33.

11b. ———. "Fior, Antonio Maria." *Dizionario biografico degli italiani*, vol. xlviii. Rome: Istituto della Enciclopedia Italiana fondata da Giovanni Treccani, 1997, pp. 92–93.

12. Berggren, J. Lennart. *Episodes in the Mathematics of Medieval Islam*, 2nd ed. New York: Springer Science+Business Media, 2016[1986].

13. Besana, Luigi. "La *Nova Scientia* di Nicolò Tartaglia: Una lettura." In Marco Berreta, Felice Mondella, and Maria Teresa Monti (eds.), *Per una storia critica della scienza*. Bologna: Cisalpino, 1996, pp. 49–71.

14. Betti, Gian Luigi. "Cardano a Bologna e la sua polemica con il Tartaglia nel ricordo di un contemporaneo." *Bruniana & Campanelliana*, n. 1, 2009, pp. 159–69.

15. Betti, Renato. "Dalle equazioni a Cardano, da Cardano all'algebra: La lunga storia delle equazioni algebriche." *Lettera matematica Pristem*, n. 41, 2001, pp. 40–45. Available at: matematica.unibocconi.it/cardano/equazioni-algebra.htm.

16. Biagioli, Mario. "The social status of Italian mathematicians, 1450–1600." *History of Science*, vol. xxvvii, 1989, pp. 41–95.

17. Bianca, Concetta. "Dal Ferro (Del Ferro), Scipione." In *Dizionario biografico degli italiani*, vol. xxxi. Rome: Istituto della Enciclopedia Italiana fondata da Giovanni Treccani, 1985, pp. 764–66.

18. Bittanti, Luigi. *Di Nicolò Tartaglia matematico bresciano*. Brescia: Apollonio, 1894.

19. Boas, Marie. *Il Rinascimento scientifico (1450–1630)*, translated by Enrico Bellone. Milan: Feltrinelli, 1973. English version: *The Scientific Renaissance 1450–1630*. New York: Harper and Brothers, 1962.

20. Bombelli, Rafael. *Opera su l'algebra (1572)*, edited by Ettore Bortolotti. Milan: Feltrinelli, 1966.

21a. Bortolotti, Ettore. "Italiani scopritori e promotori di teorie algebriche." *Annuario della Reale Università di Modena*, 1918–1919, pp. 3–102.

21b. ———. "L'algebra nella scuola matematica bolognese del secolo xvi." *Periodico di Matematiche*, s. iv, vol. v, 1925, pp. 147–85.

21c. ———. "Sulla scoperta della risoluzione algebrica delle equazioni del quarto grado." *Periodico di Matematiche*, s. iv, vol. vi, 1926, pp. 217–30.

21d. ———. "I contributi del Tartaglia, del Cardano, del Ferrari e della scuola matematica bolognese alla teoria algebrica delle equazioni cubiche." *Studi e memorie per la storia dell'Università di Bologna*, vol. ix, 1926, pp. 57–108.

21e. ———. "Le matematiche disfide, e la importanza che esse ebbero nella storia delle scienze." *Atti della Società italiana per il progresso delle scienze*, xv. Bologna: Riunione, 1927, pp. 163–80.

21f. ———. "Disputazioni matematiche nel secolo xvi." *Bollettino della Unione Matematica Italiana*, vol. vi, 1927, pp. 23–27.

21g. ———. *Studi e ricerche sulla storia della matematica in Italia nei secoli xvi e xvii*. Bologna: Zanichelli, 1928.

21h. ———. "I cartelli di matematica disfida e la personalità psichica e morale di Girolamo Cardano." *Studi e memorie per la storia dell'Università di Bologna*, vol. xii, 1935, pp. 3–79.

21i. ———. *La storia della matematica nella Università di Bologna*. Bologna: Zanichelli, 1947.

22a. Bottazzini, Umberto. "Matematica e meccanica nel Cinquecento." *Nuova civiltà delle macchine*, n. 2/3 (46–47), 1994, pp. 101–6.

22b. ———. "La "grande arte": l'algebra nel Rinascimento." In AAVV., *Storia della scienza*, under the direction of Paolo Rossi, Istituto Geografico De Agostini, Novara 2006, special edition produced by Gruppo Editoriale L'Espresso, vol. i, *La rivoluzione scientifica: Dal Rinascimento a Newton*, pp. 59–84.

23. Boyer, Carl Benjamin, and Merzbach, Uta C. *A History of Mathematics*, 2nd ed. New York: John Wiley & Sons, 1991.

24. Candido, Giacomo. "Le risoluzioni della equazione di quarto grado (Ferrari-Eulero-Lagrange)." *Periodico di Matematiche*, s. iv, vol. xxi, 1941, pp. 88–106.

25. Cantor, Moritz. "I sei cartelli di matematica disfida, primamente intorno alla generale risoluzione delle equazioni cubiche di Lodovico Ferrari coi sei controcartelli in risposta di Nicolò Tartaglia comprendenti le soluzioni de' quesiti dall'una e dall'altra parte proposti, raccolti, autografati e pubblicati da Enrico Giordani, Bolognese," translated by Alfonso Sparagna. *Bullettino Boncompagni*, vol. xi, Rome 1878, pp. 177–96.

26a. Cardano, Gerolamo. *De malo recentiorum medicorum usu medendi libellus*, apud Octavianum Scotum, Venetiis, 1536.

26b. ———. *Practica arithmetice et mensurandi singularis*. Mediolani: Antonius Castellioneus, 1539.

26c. ———. *Artis magnae, sive de regulis algebraicis*, per Johannem Petreium excusum, Norimbergae, 1545.

26d. ———. *Vita Ludovici Ferrarii Bononiensis*. In Gerolamo Cardano, *Opera Omnia*, vol. ix. Lyon: Huguetan & Ravaud, 1663, pp. 568–69.

26e. ———. *Della mia vita*, edited by Alfonso Ingegno. Milan: Serra e Riva Editori, 1982. English version: *The Book of My Life*, translated from the Latin by Jean Stoner, introduction by Anthony Grafton. New York: The New York Review of Books, 2002.

26f. ———. *Liber de ludo aleae*, edited by Massimo Tamborini. Milan: Franco Angeli, 2006.

26g. ———. *The Rules of Algebra (Ars magna)*, translated by T. Richard Witmer, with an afterword by Oystein Ore. Mineola, NY: Dover Publications, 2007.

27. Casciati, Fabio. "Gerolamo Cardano: Uno dei polymath più affascinanti nel percorso tra san Tommaso e Galileo verso la Meccanica." In Aa.Vv., *Gerolamo Cardano nel suo tempo: Atti del Convegno, 16–17 novembre 2001 (Castello Visconti di San Vito, Somma Lombardo, Varese)*. Pavia: Edizioni Cardano, 2003, pp. 11–18.

28a. Cassina, Ugo. "Risoluzione graduale dell'equazione cubica di Leonardo Pisano," *Atti della Reale Accademia delle scienze di Torino*, vol. lix, 1924, pp. 14–29.

28b. ———. *Dalla geometria egiziana alla matematica moderna*. Rome: Cremonese, 1961.

29. Ceccherini, Irene. "La genesi della scrittura mercantesca." In Otto Kresten and Franz Lackner (eds.), *Régionalisme et internationalisme: Problèmes de Paléographie et de Codicologie du Moyen Âge. Actes du xve Colloque du Comité International de Paléographie Latine (Vienna, 13–17 September 2005)*. Vienna: Verlag der Österreichischen Akademie der Wissenschaften, 2008, pp. 123–37.

30. Crossley, John N. *The Emergence of Number*. Singapore: World Scientific, 1987.

31. Cuomo, Serafina. "Niccolo Tartaglia: Mathematics, ballistics and the power of possession of knowledge." *Endeavour*, vol. xxii, n. 1, 1998, pp. 31–35.

32. Descartes, René. *Discours de la méthode pour bien conduire sa raison, et chercher la vérité dans les sciences: Plus la dioptrique, les météores, et la géométrie, qui sont des essais de cette méthode*. Leyde: Ian Maire, 1637. English version: *Discourse on Method, Optics, Geometry, and Meteorology*, rev. ed., translated, with an introduction by Paul J. Olscamp. Indianapolis and Cambridge: Hackett, 2001.

33. Ekert, Artur. "Complex and Unpredictable Cardano." *International Journal of Theoretical Physics*, vol. xlvii, n. 8, 2008, pp. 2101–19.

34. Favaro, Antonio. "Per la biografia di Niccolò Tartaglia." *Archivio storico italiano*, n. 270, a. lxxi, vol. i, dispensa 2, 1913, pp. 335–72.

35. Feldman, Richard W. "The Cardano-Tartaglia Dispute." *Mathematics Teacher*, 54, March 1961, pp. 160–63.

36. Ferrari, Ludovico, and Tartaglia, Niccolò. *Cartelli di sfida matematica*, facsimile reproduction of the original editions 1547–1548, introduction by Arnaldo Masotti. Brescia: La Nuova Cartografica, 1974.

37. Fibonacci, Leonardo. *Fibonacci's Liber abaci: A Translation into Modern English of Leonardo Pisano's* Book of Calculation, translated by Laurence Sigler. New York: Springer Science+Business Media, 2003.

38. Fierz, Markus. *Girolamo Cardano (1501–1576): Physician, Natural Philosopher, Mathematician, Astrologer, and Interpreter of Dreams*, translated from German by Helga Niman. Boston: Birkhäuser, 1983.

39. Fiocca, Alessandra. "Alcune opere inedite di Ludovico Ferrari." *Bollettino di storia delle scienze matematiche*, vol. viii, fasc. 2, 1988, pp. 239–305.

40a. Franci, Raffaella. "La rivoluzione commerciale e l'introduzione delle cifre indo-arabiche in Italia." In AA.Vv., *Storicità attualità della cultura scientifica e insegnamento delle scienze*. Casale Monferrato: Marietti, 1986, pp. 53–71.

40b. ———. "Le matematiche dell'abaco nel Quattrocento." In AAVV., *Contributi alla storia delle matematiche: Scritti in onore di Gino Arrighi*. Modena: Mucchi, 1992, pp. 53–74.

40c. ———. "La matematica dell'abaco in Italia dal xiii al xvi secolo." In Giovanni Frosali and Massimiliano Ottaviani (eds.), *Atti del Convegno "Il pensiero matematico nella ricerca storica italiana" (Ancona, 26–28 marzo 1992)*. Ancona: Tipolitografia Trifogli, 1993, pp. 62–67.

40d. ———. "La trattatistica d'abaco nel Quattrocento." In Enrico Giusti (ed.), *Luca Pacioli e la matematica del Rinascimento: Atti del convegno internazionale di studi (Sansepolcro 13–16 aprile 1994)*. Città di Castello: Petruzzi, 1998, pp. 61–75.

40e. ———. "Leonardo Pisano e la tratattistica dell'abaco in Italia nei secoli XIV e XV." *Bollettino di storia delle scienze matematiche*, vol. XXIII, fasc. 2, 2003, pp. 33–54.

41a. Franci, Raffaella, and Laura Toti Rigatelli. *Storia della teoria delle equazioni algebriche*. Milan: Mursia, 1979.

41b. ———. *Introduzione all'aritmetica mercantile del Medioevo e del Rinascimento*. Urbino: Quattro Venti, 1982.

42. Frati, Lodovico. "Scipione Dal Ferro." *Studi e memorie per la storia dell'Università di Bologna*, vol. ii, 1911, pp. 195–205.

43a. Freguglia, Paolo. "Niccolò Tartaglia e il rinnovamento delle matematiche nel Cinquecento." In AA.VV., *Cultura, scienze e tecniche nella Venezia del Cinquecento. Atti del convegno internazionale di studio: Giovan Battista Benedetti e il suo tempo*. Venice: Istituto veneto di scienze, lettere ed arti, 1987, pp. 203–16.

43b. ———. "Sur la théorie des équations algébriques entre le xvi et le xvii siècle." *Bollettino di storia delle scienze matematiche*, n. 2, 1994, pp. 259–98.

44a. Gabrieli, Giovanni Battista. *Nicolò Tartaglia: Invenzioni, disfide e sfortune*. Siena: Università degli Studi di Siena, 1986.

44b. ———. *Nicolò Tartaglia: Una vita travagliata al servizio della matematica*. Bagnolo Mella: Grafica Sette, 1997.

44c. ———. "Nicolò Tartaglia matematico bresciano." *Commentari dell'Ateneo di Brescia per l'anno 1998*. Brescia: Geroldi, 2002, pp. 323–45.

45a. Gamba, Enrico, and Montebelli, Vico. "La matematica abachistica tra ricupero della tradizione e rinnovamento scientifico." In AA.VV., *Cultura, scienze e tecniche nella Venezia del Cinquecento. Atti del convegno internazionale di studio: Giovan Battista Benedetti e il suo tempo*. Venice: Istituto veneto di scienze, lettere ed arti, 1987, pp. 169–202.

45b. ———. "Piero della Francesca matematico." *Le Scienze*, n. 331, marzo 1996, pp. 70–77.

46. Gandz, Solomon. "The Origin and Development of the Quadratic Equations in Babylonian, Greek, and Early Arabic Algebra." *Osiris*, vol. 3, 1937, pp. 405–557.

47. Garibotto, Celestino. "Scuole e maestri a Verona nel Cinquecento." *Atti e memorie dell'Accademia di agricoltura, scienze e lettere di Verona*, s. iv, vol. xxiv, 1923, pp. 195–225.

48. Garibotto, Eloisa. "Le scuole d'abbaco a Verona." *Atti e memorie dell'Accademia di agricoltura, scienze e lettere di Verona*, s. iv, vol. xxiv, 1923, pp. 315–28.

49. Gavagna, Veronica. "Alcune osservazioni sulla *Practica arithmetice* di Cardano e la tradizione abachistica quattrocentesca." In Marialuisa Baldi and Guido Canziani (eds.), *Girolamo Cardano: Le opere, le fonti, la vita*. Milan: Franco Angeli, 1999, pp. 273–312.

50a. Gherardi, Silvestro. "Di alcuni materiali per la storia della Facoltà matematica nell'antica Università di Bologna/" *Nuovi Annali delle Scienze Naturali di Bologna*, s. ii, vol. v, 1846, pp. 161–87, 244–68, 321–56, 401–36.

50b. ———. "Lettera del Prof. Silvestro Gherardi a Monsignor Gaspare Grassellini sopra alcuni cenni della vita e delle fatiche di Lodovico Ferrari, desunti dai materiali per la storia della Facoltà Matematica dell'antica Università di Bologna da Gherardi medesimo raccolti, e già comunicati in parte all'Accademia." *Nuovi annali delle scienze naturali e rendiconto delle sessioni della Società agraria e dell'Accademia delle scienze dell'Istituto di Bologna*, s. iii, vol. i, 1850, pp. 213–24.

51. Giacardi, Livia, and Roero, Clara Silvia. *La matematica delle civiltà arcaiche. Egitto, Mesopotamia, Grecia*, foreword and introduction by Tullio Viola. Turin: Stampatori, 1979.

52. Gillings, Richard J. *Mathematics in the Time of the Pharaohs*. Cambridge, MA: MIT Press, 1972. Updated edition, Mineola, NY: Dover Publications, 1982.

53. Giusti, Enrico. "Matematica e commercio nel Liber Abaci." In Enrico Giusti (ed., in collaboration with Raffaella Petti), *Un ponte sul Mediterraneo: Leonardo Pisano, la scienza araba e la rinascita della matematica in Occidente*. Florence: Polistampa, 2002, pp. 59–120. Available at: php.math.unifi.it/archimede/archimede/fibonacci/catalogo/giusti.php.

54. Gliozzi, Giuliano. "Cardano, Gerolamo." In *Dizionario biografico degli italiani*. Rome: Istituto della Enciclopedia Italiana fondata da Giovanni Treccani, vol. xix, 1976, pp. 758–63.

55. Gliozzi, Mario. "Cardano, Gerolamo." In Charles Coulston Gillispie (ed.), *Dictionary of Scientific Biography*, vol. iii. New York: Charles Scribner's Sons, 1971, pp. 64–67.

56. Grendler, Paul F. *Schooling in Renaissance Italy: Literacy and Learning, 1300–1600*. Baltimore and London: Johns Hopkins University Press, 1991.

57. Heath, Thomas Little. *Diophantus of Alexandria. A Study in the History of Greek Algebra. With a supplement containing an account of Fermat's theorems and problems connected with Diophantine analysis and some solutions of Diophantine problems by Euler.* Mineola, NY: Dover Publications, 1964.

58. Helbing, Mario Otto. "La scienza della meccanica nel Cinquecento." In Antonio Clericuzio and and Germana Ernst (eds., in collaboration with Maria Conforti). *Il Rinascimento italiano e l'Europa*, vol. v: *Le scienze*. Treviso-Costabissara: Angelo Colla, 2008, pp. 573–92.

59. Jayawardene, Sugathadasa A. "Ferrari, Ludovico." In Charles Coulston Gillispie (ed.), *Dictionary of Scientific Biography*. New York: Charles Scribner's Sons, vol. iv, 1971, pp. 586–88.

60. Joseph, George Gheverghese. *C'era una volta un numero*, translated by Barbara Mussini. Milan: Il Saggiatore, 2000. English version: *The Crest of the Peacock. Non-European Roots of Mathematics*, 3rd ed. Princeton and Oxford: Princeton University Press, 2011.

61. Khayyam, Omar. *L'algèbre d'Omar AlKayyani*, edited by Franz Woepke. Paris: Duprat, 1851. English version: *The Algebra of Omar Khayyam*, edited by Daoud S. Kasir. New York: Bureau of Publications, Teachers College, Columbia University, 1931.

62a. Khwarizmi, Muhammad ibn Musa al-. *Algoritmi de numero indorum*. Rome: Baldassarre Boncompagni, Tipografia delle scienze fisiche e matematiche, 1857.

62b. ———. *The Book of Algebra*, translated by Frederick Rosen. Islamabad: Pakistan Hijra Council, 1989.

63. Kline, Morris. *Mathematical Thought from Ancient to Modern Times*, vol. 1. New York and Oxford: Oxford University Press, 1972.

64. Li Vigni, Ida. "Gerolamo Cardano: l'autobiografia come mito di sé." *Anthropos & Iatria*, nn. 2–3, 1998, pp. 55–59.

65. Long, Pamela O. *Openness, Secrecy, Authorship: Technical Arts and the Culture of Knowledge from Antiquity to the Renaissance*. Baltimore and London: Johns Hopkins University Press, 2001.

66a. Loria, Gino. *Le scienze esatte nell'antica Grecia*. Milan: Hoepli, 1914.

66b. ———. *La storia delle matematiche dall'alba delle civiltà al tramonto del secolo xix*. Milan: Hoepli, 1950.

67. Maccagni, Carlo. "La stampa, le scienze e le tecniche a Venezia e nel Veneto tra Quattrocento e Cinquecento." In AA.VV., *Cultura, scienze e tecniche nella Venezia del Cinquecento. Atti del convegno internazionale di studio: Giovan Battista Benedetti e il suo tempo*. Venice: Istituto veneto di scienze, lettere ed arti, 1987, pp. 483–94.

68. Maiocchi, Roberto. *Storia della scienza in Occidente: Dalle origini alla bomba atomica*. Florence: La Nuova Italia, 2000.

69. Manzoni, Dominico. *Quaderno doppio col suo giornale, novamente composto et diligentissimamente ordinato secondo il costume di Venetia: Opera a ogni persona utilissima, et molto necessaria*, per Comin da Trino di Monferrato, Venice, 1554.

70a. Maracchia, Silvio. *Da Cardano a Galois: Momenti di storia dell'algebra*. Milan: Feltrinelli, 1979.

70b. ———. "Cardano e la storia della matematica." In Oscar Montalto and Lucia Grugnetti (eds.), *La storia delle matematiche in Italia. Atti del Convegno: Cagliari, 29–30 settembre e 1 ottobre 1982*. Cagliari: Università Istituti di matematica delle facoltà di scienze e ingegneria, 1983, pp. 351–57.

70c. ———. "Tartaglia e la sua 'regola generale'." *Commentari dell'Ateneo di Brescia per l'anno 1998*. Brescia: Geroldi, 2002, pp. 275–87.

70d. ———. "Algebra e geometria in Cardano." In Marialuisa Baldi and Guido Canziani (eds.), *Cardano e la tradizione dei saperi. Atti del Convegno internazionale di studi, Milano, 23–25 maggio 2002*. Milan: Franco Angeli, 2003, pp. 145–55.

70e. ———. *Storia dell'algebra*. Naples: Liguori, 2005.

71a. Masotti, Arnaldo. *Commemorazione di Niccolò Tartaglia*. Brescia: Geroldi, 1958.

71b. ———. "Sui Cartelli di matematica disfida scambiati fra Lodovico Ferrari e Niccolò Tartaglia." *Rendiconti dell'Istituto Lombardo, Accademia di Scienze e Lettere*, vol. 94, 1960, pp. 31–41.

71c. ——— (ed.), *Atti del convegno di storia delle matematiche, 30–31 maggio 1959: quarto centenario della morte di Niccolò Tartaglia*. Brescia: La Nuova Cartografica, 1962.

71d. ———. "Ferro (or Ferreo, Dal Ferro, Del Ferro), Scipione." In Charles Coulston Gillispie (ed.), *Dictionary of Scientific Biography*. New York: Charles Scribner's Sons, vol. 4, 1971, pp. 595–97.

71e. ———. "Gabriele Tadino e Niccolò Tartaglia." *Atti dell'Ateneo di scienze, lettere e arti di Bergamo*, vol. 38, 1973–74, pp. 361–74.

71f. ———. "Tartaglia, Niccolò." In Charles Coulston Gillispie (ed.), *Dictionary of Scientific Biography*, vol. 13. New York: Charles Scribner's Sons, 1976, pp. 258–62.

72. Matthiae, Paolo. *I tesori di Ebla*. Rome and Bari: Laterza, 1984.

73. Mazzetti, Serafino. *Repertorio dei professori dell'Universitàe dell'Istituto delle scienze di Bologna*. Sala Bolognese: Forni, 1988 (reprint of the edition: Tipografia di san Tommaso d'Aquino, Bologna 1848).

74. Milani, Mino. *Gerolamo Cardano: Mistero e scienza nel Cinquecento*. Milan: Camunia, 1990.

75a. Montebelli, Vico. "Ex falsis verum: il metodo della falsa posizione, semplice e doppia, nell'ambito della matematica abachistica del Medioevo e del Rinascimento." *Quaderni dell'Accademia fanestre*, n. 3, 2004, pp. 191–230.

75b. ———. "Alle origini della matematica applicata: Le scuole d'abaco." Available at: www2 .polito.it/didattica/polymath/htmlS/Interventi/DOCUMENT/IdroUno/Media /Montebelli%20-%20Alle%20origini%20della%20matematica%20applicata_le%20 scuole%20d'abaco.pdf.

76. Morley, Henry. *Jerome Cardan: The Life of Girolamo Cardano, of Milan, Physician*. London: Chapman and Hall, 1854.

77. Nenci, Elio. "Le ricerche matematiche tra segretezza e pubbliche dispute." In Antonio Clericuzio and Germana Ernst (eds., in collaboration with Maria Conforti), *Il Rinascimento italiano e l'Europa*, vol. v: *Le scienze*. Treviso-Costabissara: Angelo Colla, 2008, pp. 627–40.

78. Neugebauer, Otto. *Le scienze esatte nell'antichità*, translated by Libero Sosio. Milan: Feltrinelli, 1974. English version: *The Exact Sciences in Antiquity*, 2nd ed. Mineola, NY: Dover Publications, 1969.

79. Nordgaard, Martin A. "Sidelights on the Cardan-Tartaglia Controversy." *National Mathematics Magazine*, vol. xii, n. 7, April 1938, pp. 327–46.

80. Notari, Vittoria. "L'equazione di quarto grado: Risoluzione di Lodovico Ferrari e sua inter-pretazione geometrica." *Periodico di Matematiche*, s. iv, vol. iv, 1924, pp. 327–34.

81. Ore, Oystein. *Cardano: The Gambling Scholar.* Princeton, NJ: Princeton University Press, 1953.

82. Pacioli, Luca. *Summa de arithmetica, geometria, proportioni et proportionalità.* Venice: Paganino de Paganini, 1494.

83a. Pascuale, Luigi di. "Le equazioni di terzo grado nei *Quesiti et inventioni diverse* di Nicolò Tartaglia." *Periodico di Matematiche*, s. iv, vol. xxxv, 1957, pp. 79–93.

83b. ———. "I cartelli di matematica disfida di Ludovico Ferrari e i controcartelli di Nicolò Tartaglia," parte i. *Periodico di Matematiche*, s. iv, vol. xxxv, 1957, pp. 253–78.

83c. ———. "I cartelli di matematica disfida di Ludovico Ferrari e i controcartelli di Nicolò Tartaglia," parte ii. *Periodico di Matematiche*, s. iv, vol. xxxvi, 1958, pp. 175–98.

84a. Pettinato, Giovanni. *Ebla, nuovi orizzonti della storia.* Milan: Rusconi, 1986.

84b. ———. *La città sepolta: I misteri di Ebla.* Milan: Mondadori, 1999.

85. Piotti, Mario. *La lingua di Niccolò Tartaglia: La "Nova Scientia" e i "Quesiti et inventioni diverse."* Milan: Edizioni Universitarie di Lettere Economia Diritto (LED), 1998.

86a. Pizzamiglio, Pierluigi. "L'Ateneo di Brescia e Niccolò Tartaglia." In AA.VV., *L'Ateneo di Brescia e la storia della scienza.* Brescia: Geroldi, 1985, pp. 87–105.

86b. ———. "Gerolamo Cardano (1501–1576)." *L'insegnamento della matematica e delle scienze integrate*, vol. xxiv, sez. A, May 2001, pp. 238–60.

86c. ———. "Niccolò Tartaglia (1500 ca.–1557) nella storiografia." *Memorie scientifiche, giuridiche, letterarie*, s. viii, vol. viii, fasc. 2, 2005, pp. 443–53.

87. Procissi, Angiolo. "Il caso irriducibile dell'equazione cubica da Cardano ai moderni alge-bristi." *Periodico di Matematiche*, s. iv, vol. xxix, 1951, pp. 263–80.

88. Rashed, Roshdi. *The Development of Arabic Mathematics: Between Arithmetic and Algebra,* translated by Angela F. W. Armstrong. Dordrecht: Kluwer, 1994.

89. Rashed, Roshdi, and Vahabzadeh, Bijan. *Al-Khayyam mathématicien.* Paris: Blanchard, 1999.

90. Roero, Clara Silvia. "Algebra e aritmetica nel Medioevo islamico." In Enrico Giusti (ed., in collaboration with Raffaella Petti), *Un ponte sul Mediterraneo: Leonardo Pisano, la scienza araba e la rinascita della matematica in Occidente.* Florence: Polistampa, 2002, pp. 7–43. Avail-able at: php.math.unifi.it/archimede/archimede/fibonacci/catalogo/roero.php.

91. Rose, Paul Lawrence. *The Italian Renaissance of Mathematics: Studies on Humanists and Math-ematicians from Petrarch to Galileo.* Geneva: Librairie Droz, 1975.

92. Schwarz, Matthäus. "Copia und Abschrift ab und von Matheus Schwartzen aigen Hand-schrift, was das Buchhalten sey; auch von dreierlay Buchhalten, so er inn seiner Jugent also selbst gestalt und gemacht hat, als im 1516 und 1518." In Alfred Weitnauer, *Venezianischer Handel der Fugger nach der Musterbuchhaltung des Matthäus Schwarz.* Munich and Leipzig: Duncker und Humblot, 1931, pp. 174–314.

93. Siraisi, Nancy Gillian. *The Clock and the Mirror: Girolamo Cardano and Renaissance Medicine.* Princeton, NJ: Princeton University Press, 1997.

94. Smith, David Eugene, and Karspinki,, Louis Charles. *The Hindu-Arabic Numerals.* Boston and London: Ginn and Company Publishers, 1911.

95. Stewart, Ian. *Why Beauty Is Truth: A History of Symmetry.* New York: Basic Books, 2007.

96. Struik, Dirk Jan. *Matematica: Un profilo storico*, translated by Umberto Bottazzini and Virginio Sella, with an appendix by Umberto Bottazzini. Bologna: il Mulino, 1981. English version: *A Concise History of Mathematics*, 4th rev. ed., Mineola, NY: Dover Publications, 1987.

97. Tadini, Guido. *Gabriele Tadino priore di Barletta*. Bergamo: Bolis, 1986.

98. Tamborini, Massimo. *De cubo et rebus aequalibus numero: La genesi del metodo analitico nella teoria delle equazioni cubiche di Girolamo Cardano*. Milan: Franco Angeli, 1999.

99. Tanner-Young, Rosalind C. H. "The alien realm of the minus: deviatory mathematics in Cardano writings." *Annals of Science*, vol. xxxvii, 1980, pp. 159–78.

100a. Tartaglia, Niccolò *Nova scientia inventa da Nicolo Tartalea*, per Stephano da Sabio ad instantia di Nicolo Tartalea, in Vinegia, 1537.

100b. ———. *La nova scientia de Nicolo Tartaglia: Con una gionta al terzo libro*, [s.n.], in Vinegia 1558.

100c. ———. *Quesiti et inventioni diverse*, facsimile reproduction of the 1554 edition with an introduction by Arnaldo Masotti. Brescia: La Nuova Cartografica, 1959.

100d. ———. *General trattato di numeri, et misure*, 6 parti, Curzio Troiano de' Navò, Venice, 1556–1560.

101. Tignol, Jean-Pierre. *Galois' Theory of Algebraic Equations*. Singapore: World Scientific, 2001.

102. Toti Rigatelli, Laura. "Documenti per una storia dell'algebra in Italia dal XIII al XVI secolo." In Oscar Montalto and Lucia Grugnetti (eds.), *La storia delle matematiche in Italia. Atti del Convegno (Cagliari, 29–30 September and 1 October 1982)*. Istituti di matematica delle facoltà di scienze e ingegneria, Università di Cagliari 1983, pp. 333–39.

103a. Ulivi, Elisabetta. "Scuole e maestri d'abaco in Italia tra Medioevo e Rinascimento." In Enrico Giusti (ed., in collaboration with Raffaella Petti), *Un ponte sul Mediterraneo: Leonardo Pisano, la scienza araba e la rinascita della matematica in Occidente*. Florence: Polistampa, 2002, pp. 121–55. Available at: php.math.unifi.it/archimede/archimede/fibonacci/catalogo/ulivi.php.

103b. ———. "Scuole d'abaco e insegnamento della matematica." In Antonio Clericuzio and Germana Ernst (eds., in collaboration with Maria Conforti), *Il Rinascimento italiano e l'Europa*, vol. v: *Le scienze*. Treviso-Costabissara: Angelo Colla, 2008, pp. 403–20.

103c. ———. "Su Leonardo Fibonacci e i suoi maestri d'abaco pisani dei secoli XIII–XV." *Bollettino di storia delle scienze matematiche*, vol. XXXI, fasc. 2, 2011, pp. 247–86.

104. Villa, Mario. "Niccolò Tartaglia a quattro secoli dalla morte." *Atti della Accademia delle Scienze dell'Istituto di Bologna, Classe di Scienze Fisiche: Rendiconti*, s. xii, vol. i, 1963–1964, pp. 5–24.

105. Vino, Isabella, and Viola, Tullio. "Un problema algebrico." In *Testi lessicali monolingui*, appendice D., testo n. 73 della Biblioteca L. 2769. Naples: Istituto Universitario Orientale di Napoli, Seminario di studi asiatici, Series maior, iii, Materiali epigrafici di Ebla (MEE) 3, 1981, pp. 278–85.

106a. Waerden, Bartel Leender van der. *Science Awakening*, English translation by Arnold Dresden. New York: Oxford University Press, 1961.

106b. ———. *A History of Algebra from Al-Khwarizmi to Emmy Noether*. New York: Springer Verlag, 1985.

107. Ziliani, Carolina. "Note su la vita e sull'opera di Niccolò Tartaglia." *Studium*, n. 2, February 1926, pp. 84–92.

ADDITIONAL BIBLIOGRAPHY
FOR THE ENGLISH EDITION

Black, Robert. *Humanism and Education in Medieval and Renaissance Italy: Tradition and Innovation in Latin Schools from the Twelfth to the Fifteenth Century.* Cambridge: Cambridge University Press, 2001.

Crossley, John N., and Henry, Alan S. "Thus spake al-Khwarizmi: A translation of the text of Cambridge University Library Ms. li.vi.5." *Historia Mathematica*, vol. 17, 1990, pp. 103–31.

Høyrup, Jens. "Leonardo Fibonacci and abbaco culture: A proposal to invert the roles." *Revue d'histoire des mathématiques*, vol. 11, 2005, pp. 23–56.

———. *Lengths, Widths, Surfaces. A Portrait of Old Babylonian Algebra and Its Kin.* Berlin and Heidelberg: Springer, 2002.

Imhausen, Annette. *Mathematics in Ancient Egypt: A Contextual History.* Princeton and Oxford: Princeton University Press, 2016.

Katz, Victor J., and Parshall, Karen Hunger. *Taming the Unknown: A History of Algebra from Antiquity to the Early Twentieth Century.* Princeton and Oxford: Princeton University Press, 2014.

Oaks, Jeffrey A., and Alkhateeb, Haitham M. "Simplifying equations in Arabic algebra." *Historia Mathematica*, vol. 34, n. 1, February 2007, pp. 45–61.

Robson, Eleanor. *Mathematics in Ancient Iraq: A Social History.* Princeton and Oxford: Princeton University Press, 2008.

Sesiano, Jacques. *An Introduction to the History of Algebra: Solving Equations from Mesopotamian Times to the Renaissance,* translated by Anna Pierrehumbert. Providence, RI: American Mathematical Society, 2009.

Stedall, Jacqueline. *From Cardano's Great Art to Lagrange's Reflections: Filling a Gap in the History of Algebra.* Zürich: European Mathematical Society, 2011.

Ulivi, Elisabetta. "Masters, questions and challenges in the abacus schools." *Archive for History of Exact Sciences,* vol. 69, no. 6 (November 2015), pp. 651–70.

INDEX

A NOTE ON THE TYPE

This book has been composed in Arno, an Old-style serif typeface in the classic Venetian tradition, designed by Robert Slimbach at Adobe.